Ethics of Virtual Reality

Ethics of Virtual Reality

Technology and Experience

Michael J. Grabowski

LEXINGTON BOOKS
Lanham • Boulder • New York • London

Published by Lexington Books
An imprint of The Rowman & Littlefield Publishing Group, Inc.
4501 Forbes Boulevard, Suite 200, Lanham, Maryland 20706
www.rowman.com

86-90 Paul Street, London EC2A 4NE, United Kingdom

British Library Cataloguing in Publication Information Available

Library of Congress Cataloging-in-Publication Data Available

ISBN 9781793651976 (cloth) | ISBN 9781793651983 (epub)

♾™ The paper used in this publication meets the minimum requirements of American National Standard for Information Sciences—Permanence of Paper for Printed Library Materials, ANSI/NISO Z39.48-1992.

Contents

List of Figures

Acknowledgments

I would like to thank the many people who made this book possible. First, I am incredibly grateful to my wife Laura, who has read every word of this book and offered invaluable feedback that has made it better. I thank my colleagues in my department and elsewhere at Manhattan University for allowing me to engage in endless discussions about virtual reality, artificial intelligence, and the need to not lose sight of human experience while immersed in technology. I have tested many of these ideas at scholarly conferences such as the Media Ecology Association, Society for the Cognitive Studies of the Moving Image, the Institute for General Semantics, and the University Film and Video Association. Thank you to the members of each group for helping me refine my thinking and offering feedback and encouragement. I would also like to thank my students, particularly in my VR Perception and Production course, for teaching me a thing or two while we explored how we sense natural and artificial environments and how we can fool our senses using the immersive environments that VR generates.

Thank you to everyone at Lexington Books. I owe a debt of gratitude to my editor, Jessie Tepper, for having an abundance of patience. I also thank Charles Tropp for his detail in producing a far more useful index of the book than I ever could. As I worked my way through each chapter, I came to realize that there is one dimension that virtual reality cannot easily reproduce: time. The time I spent writing this book was time I did not spend with my children, though they often were nearby as I typed. The future I describe in these pages is the one they will inherit, and they will have to face the ethical questions that emerge as we conduct more affairs in virtual environments. I dedicate this book to Ethan, Maya, and Gillian in the hope that it provides some useful guidance for the decisions they will have to make in the world to come.

Introduction

Imagine you are flying through a virtual streetscape toward the digital building where you attend your college course. Theoretically, you could teleport directly to your virtual classroom, but the free headset you have been issued (or, in this future scenario, glasses, contact lenses, or possibly implants) requires you to traverse a commercial thoroughfare first. On your way to class, avatars approach you, hawking everything from study aids to credit card offers to digital wardrobes you can purchase to distinguish your online avatar. Reminders from your social media accounts ping in front of you—your friend from Taiwan needs another team member for virtual reality (VR) Street Brawl, while your favorite VR creator has just posted a new 360-degree video.

Although you don't really know who is a real person behind another headset and who is a bot (computer code communicating as if it were human), you have gotten pretty good at telling them apart. Sometimes, though, you are surprised—you agree to meet a virtual friend after class only to find out after chatting with her that she is an ad created to sell a soft drink brand. A few times, you have mistakenly assumed someone was a bot until you met them IRL (in real life).

You silence a notification that flashes about an exclusive VR party, sponsored by the headset manufacturer. You will be penalized for arriving late to your virtual class session, so you suppress the urge to have more fun by attending the party and earning credits to purchase a new VR app. After all, a skill you are trying to master is ignoring these digital distractions and focusing on your work, even though the companies behind these distractions invest millions of dollars refining their algorithms to break down your resistance to them. It's enough to make you want to take off the headset and retreat to the natural world, but the only way to access these classes, and eventually the credentialing and degree to allow you to work and advance in this world, is to participate in the virtual landscape—and it is likely that the best paying jobs after you graduate will be in this area.

This scenario is a possible future for VR: a dystopian scene in which a monopolistic corporation positions users to receive a barrage of ads and upsells of in-app purchases, all while enriching a few at the expense of many. It is sustained by a closed platform to which competitors cannot have access, and if any start-ups develop a competing platform that manages to peel away a small part of the market share, the dominant company simply buys them out and incorporates their project into its platform. The system maintains its control by offering users a small piece of the pie, promising them upgrades and other virtual compensation in exchange for an investment of time, money, and social connections. Some are rewarded more than others: younger and more attractive personas are viewed as more desirable. Users respond by sculpting their digital avatars and personas to conform to a perceived idealized look more Hollywood than global village. Even more troubling, some people are excluded entirely from the platform because they are considered to be a "bad investment."

Given the conditions of such a dystopic environment, it is natural to ask why anyone would choose to participate in it. For many people, spending time in a virtual world may be more desirable and fulfilling, despite its many flaws. Environmental degradation and extreme weather events, threats from mass shootings and biological dangers, and the escalating cost of housing and travel present risks in the real world that are not present in the virtual one. In VR, everyone can see the Louvre, the bottom of the ocean, or some far-off planet without leaving their home. No matter how dystopic a VR future may be, the real world may be even worse in comparison, or the company supplying the VR environment may have successfully addicted users to their service.

However, VR does not necessarily have to become a choice between the lesser of two evils. Experiencing VR environments can be helpful and a lot of fun. One of the most popular VR apps today is Supernatural, a fitness app that uses first-person views of personal trainers, breathtaking landscapes, and fun activities impossible to do in real life to motivate users during their workouts.[1] Emergency workers can learn dangerous tasks more effectively with safe VR simulations, and surgeons can plan out operations in 3D using a patient's medical data imported into VR. Psychologists have used VR to conduct exposure therapy with patients, and students can visualize historical or other inaccessible environments.[2] Apple has introduced the concept of spatial computing with its Vision Pro headset, allowing users to seamlessly transition from full VR to an augmented reality (AR) virtual desktop displayed over a view of the real world.[3] 360-degree video in VR provides an immersive experience, placing viewers in a scene rather than viewing it in a frame. People are exploring new forms of creativity with the digital music apps Tribe XR and SynthVR and digital painting and sculpture with Tilt Brush and Gravity Sketch. Fans of Vincent Van Gogh's paintings can step inside them with "The Night Café," and "Traveling While Black" educates viewers about the historical restrictions of travel for Black Americans.[4] Despite the threats of VR becoming an oppressive technology, it also offers promise of becoming a liberating one. More likely, VR will exhibit some characteristics of both as it becomes more widely adopted.

An ethical approach to studying VR can guide the direction of its development, leading to recommended uses and limitations to encourage healthy practices and provide early warnings about potential abuses of the technology. Ethics provides a framework for understanding how to put VR into practice, consider who is helped or harmed, and who should decide whether and how to employ the technology in different contexts. Beyond simply assessing VR as "good" or "evil," an ethics of VR is a reflective process that foregrounds its biases, risks, and cascading effects so that people can anticipate and prepare for the changes it brings. Ultimately, the ethics of VR helps to give people some degree of control over its development and use, rather than ceding control to others or the technology itself.

This book examines VR from several ethical perspectives to provide multiple ways of thinking about how the medium will impact us. While it relies on foundational theories of ethics, it also incorporates applied ethics of other digital technologies, including social media, digital video, and online communication. Cultures have experienced rapid technological change over the last century, affecting how people work, travel, communicate, and play. VR relies on many of these technologies to support it, but it also possesses unique characteristics that require a separate examination of its ethical development and implementation. Moreover, because VR can be used for so many different purposes, no one ethical theory can address all the possible uses to which it can be put. In the chapters that follow, several ethical approaches are applied to both real and imagined uses of VR to address questions like how to balance safety and freedom in VR, who should control or have access to VR platforms, and what kind of communities should be encouraged or suppressed in virtual worlds. While it may be difficult to determine outright solutions to address these concerns, applying established and emerging ethical frameworks can give space for reflective thinking about these issues.

IMAGINING VR

Preparing for the impacts of a new technology requires imagining what it may become and how we will use it. Even though VR is at present a developing technology, it is growing quickly and is likely to become a major form of online interaction. Over 10 million VR headsets were sold in 2021, and VR is expected to grow to become an $87 billion industry by 2030.[5] However, knowing that VR will become a mainstream technology does not necessarily explain how it will develop or how people will interact with it.

Fictional depictions of future technologies may offer insight into their possible uses before they are implemented. Just like *Star Trek* communicators and tricorders predated the iPhone, fictional depictions of VR may provide some insights into what VR may become, as well as the social anxiety generated over their use. Though science fiction often presents a coded form of social critique of the present, the conditions within which those critiques are made do not disappear with the emergence of a new technology—rather, they often become more acute with the acceleration of technological change. Therefore, it is useful to examine briefly how popular fiction has imagined what VR would be and how it would be used.

The blockbuster film trilogy *The Matrix* presents a virtual world as the perceived reality for most of humanity. The first film of the trilogy, released in 1999, imagines a future where machines use humans for their bioenergy and keep them alive by generating an entire digital world that most humans believe is the real one. Except for a select few liberated resistors, who "jack in" to the virtual world via a neuro-interface through their skulls, most people spend their entire lives in the Matrix and believe they exist in the late twentieth century.

The Matrix creators, the Wachowskis, imagined a dystopic future in which machines possessing artificial intelligence (AI) go to war against their human creators, like a golem or Frankenstein's monster, but they also critiqued the influence of technology more broadly, including our tendency to behave like machines to function within technocratic systems and our propensity to be satisfied by entertainment rather than addressing social and environmental problems. In the film, the character Cypher makes a deal with the AI-generated Agent Smith to be placed back into the virtual world with no memory of living outside of it in exchange for selling out the film's hero, Neo. "I know this steak doesn't exist," Cypher says while eating a computer-generated steak. "I know that when I put it in my mouth, the Matrix is telling my brain that it is juicy and delicious. After nine years, you know what I realize? Ignorance is bliss." By the end of the film, Neo can see the underlying computer code that makes up the virtual world. *The Matrix* continues to carry such a strong resonance with VR that the Unreal Engine, a major VR developer platform, distributed the cityscape from *The Matrix Resurrections*, the fourth film in the franchise, as a default VR environment.[6] (See figure 0.1.)

Long after this film's release, virtual worlds continue to face criticisms about addiction, from teen depression and suicides linked to social media

Figure 0.1. ***The Matrix*: Neo Sees the Underlying Code of the Matrix.** *Screenshot by author.*

addiction to extreme cases of multiuser video game players dying from lack of sleep.[7] These criticisms take the form of two different arguments: (1) like the Matrix, these virtual worlds are so compelling to users that they choose virtual worlds at the expense of their real-world existence, and (2) these virtual worlds are designed explicitly to addict users to their platforms. The problem is so widespread that a new market has emerged for screen time apps that help users limit the time they spend online. Later in this book, questions are posed about the design of VR platforms: As these worlds compete with one another for our attention, how do we prevent them from treating the real world as just another competitor to overcome, or should we?

In *The Matrix*, the virtual world is so real that most characters in it do not realize they are in a simulation. However, VR in real life exists more as a blend between new forms of fiction and an augmentation of reality than a replacement for reality. There may be moments when users are so immersed in a VR experience that they momentarily forget it is an illusion, but they know on a fundamental level that they are interacting within a mediated environment. Like stories about Parisian audiences who jumped out of their seats while watching *L'arrivée d'un train en gare de La Ciotat*, one of the first silent films by the Lumière Brothers of a train heading toward the camera as it pulls into the station, we may react perceptually to VR experiences as if they were real, but, unlike characters in *The Matrix*, we know they are not.[8] VR may be more immersive than previous forms of media, but we do not mistake this world for the one we live in once we take off the headset.

However, when we are communicating with other real people in a VR environment, like in a VR collaborative workspace or multiplayer game, that interaction is real even though the interface is artificial. We do not think of a telephone or Zoom call with a friend or family member as fake; instead, it is a way to communicate with someone when engaging in an in-person conversation would be inconvenient or impossible. Asking only what is real in VR may obscure more complex questions about how our real conversations are shaped in this artificial environment.

Almost 20 years after *The Matrix*, the Steven Spielberg film *Ready Player One* presented VR as "a place where the limits of reality are your own imagination," according to its teen hero, Wade Watts. The film is based on the Ernest Cline best-selling novel in which Watts competes in a contest to win control of the virtual world called OASIS (Ontologically Anthropocentric Sensory Immersive Simulation). In this story, the corporate conglomerate Innovative Online Industries (IOI) competes for control of OASIS so it can achieve its goals of profit and market dominance. OASIS' late founder James Halliday echoes Apple's Steve Jobs, an innovator famed for introducing digital technologies that have created new forms of virtual economies and social interactions. Likewise, OASIS co-founder Ogden Morrow resembles Apple's Steve Wozniak, who left the company after objecting to the direction of its vision. In Cline's novel, Morrow notes that OASIS "had become a self-imposed prison for humanity."[9]

Even so, the OASIS is presented as an escape and playground for the people who inhabit it. It is where Watts attends school, plays video games, hangs out with friends, and conducts research into 1980s trivia and pop culture to compete in the contest for control of the platform. Connecting to the OASIS from his hideout inside the back of an abandoned van, Wade can experience a far richer existence than in the trailer park stacks, where he sleeps in his aunt's laundry room. As a poor teen in real life, Watts becomes a world-famous celebrity in the virtual world, and a better-looking one, as well. He narrates, "My avatar had a slightly smaller nose than me, and he was taller. And thinner. And more muscular. And he didn't have any teenage acne." Another character dryly states, "The OASIS lets you be whoever you want to be. That's why everyone is addicted to it." (See figure 0.2.)

Ready Player One may have been an inspiration for Meta (formerly Facebook) founder Mark Zuckerberg, who renamed his company in 2021 to Meta and directed Facebook Reality Labs to build the metaverse, a virtual world where people can collaborate, have fun, and interact with family and friends around the world. More likely, Zuckerberg borrowed the name from Neal Stephenson's 1992 cyberpunk novel *Snow Crash*, where people accessed the Metaverse with a computer beaming 3D images into goggles and stereo digital audio into earphones. In Stephenson's version of VR, the Metaverse resembled an urban center called the Street. "It does not really exist," he writes. "But right now, millions of people are walking up and down it."[10]

Zuckerberg told investors,

Figure 0.2. The Hideout Where Wade Connects to OASIS. *Screenshot by author.*

> We believe the Metaverse will be the successor to the mobile internet. . . . We'll be able to feel present like we are right there with people no matter how far apart we actually are. We'll be able to express ourselves in new joyful, completely immersive ways.[11]

However, critics worry that the metaverse will follow the same path as Meta's other products, which sell users' data to advertisers and have been subjected to criticism for privacy concerns, misinformation, and political propaganda. Like Watts' warning about IOI gaining control of OASIS, advocates of an open-source VR seek privacy protections and interoperability between headsets to prevent a private monopoly of the virtual world. When Facebook, before it changed its name to Meta, announced that owners of the Oculus Quest would require a Facebook account in order to access it, open-source activists protested and began a project to build a non-branded headset.[12]

Meta has since indicated that the metaverse will be decentralized and open to multiple platforms. They perhaps are motivated by Apple's release of its own headset, called Vision Pro, which combines the ability to integrate digital graphics with a view of the physical world (AR) and digital graphics that obscure the view of the physical world (VR).[13] Apple hired Hollywood directors like Jon Favreau and partnered with Disney to create immersive content for this new headset, signaling that it will embrace storytelling as a central feature, but its integration of AR features suggests that it may also go after Microsoft's HoloLens, an AR visor that the company claims is designed "for precise, efficient hands-free work."[14] Apple's Vision Pro calls its efforts "spatial computing," defined as using the space around users to augment or replace the view and sounds of their environment, where they can interact with virtual objects, scenes, images, and data.[15]

Over the past few decades, other films and television programs have depicted people entering digital virtual worlds. In the late 1980s, *Star Trek: The Next Generation* introduced viewers to the holodeck, a room that displayed a grid pattern when it was turned off. When it was switched on, however, characters were entertained or worked out problems in an immersive and interactive three-dimensional scene, from a Western setting to the 1930s gangster genre to Victorian England to an alternative version of the ship itself.[16] Several episodes of the Netflix series *Black Mirror* used variations of VR and virtual worlds, including the episode "San Junipero," where elderly characters can live a nostalgic 1980s experience in younger bodies, even beyond their deaths.[17] Both films *Brainstorm* and *Until the End of the World* show people using machine interfaces to relive other people's experiences and dreams, and VR plays a central role in transforming a developmentally impaired man into a psychotic genius in *The Lawnmower Man.* James Cameron's *Avatar* features a paraplegic soldier who operates a genetically engineered alien body with a virtual interface, and Leonardo DiCaprio enters the dreams of others in Christopher Nolan's *Inception*, but he may be living in a virtual dreamworld himself. Before *The Matrix*, Keanu Reeves played a character with a digital storage brain implant who uses a visor and gloves to access a virtual world in *Johnny Mnemonic*.

Each of these fiction examples, among many others, imagines VR as a powerful and mesmerizing tool but also something that is imbued with danger or the capacity to be abused. Long before VR has become widely adopted, writers and directors have given thought, sometimes carefully and other times sensationally, to the different forms VR may take and the consequences it enables. By helping to develop a popular ethics of technology within dramatic structures, these stories help audiences think about how they may use VR in their lives and how it could be used against them.

VR BEFORE VR

Experiencing VR for the first time is (literally and figuratively) an eye-opening experience. Placing a Meta Quest visor over your eyes, you are first enveloped in darkness before the screen lights up with text seemingly floating in space. Before long, a home screen appears—it looks like a home—and you seem to be standing in a living room suspended above a forest, with other homes in the distance in view through your room's glass-paneled walls. Your environment appears in 3D, and you are enveloped in sound. A screen suspends in front of you, and you can use the controllers in your hand or your hands themselves to select the VR experience you want to enter. The first time someone tries this, they are amazed by the immersive experience and look around in all directions. However, most people have some idea of what to expect because previously existing media have provided the metaphors for experiences in this new environment. In addition to television shows and movies showing images of VR in use, viewers' own experiences with video games, social media, mobile computing, online video, and other digital technologies have prepared them for using VR, just as those technologies have helped VR developers imagine what experiences to build in VR. (See figure 0.3.)

One of the most obvious transfers from older media to VR is the ability to watch movies and television shows in VR. Major streaming services like YouTube, Netflix, and Amazon Prime all have VR apps to watch videos on a virtual large screen. An obvious question is: Why would anyone need to watch a show in VR when they can simply turn on a TV? The advantages of watching video in VR are that it completely envelops your experience: you are transported to a virtual room where a large screen appears in front of you. You only need a small portable headset, not a wall-sized TV, for this experience, which is a major convenience for people in small spaces. You can also filter out any distractions from the physical space in which you are sitting. Watching video in VR is like having your own private screening room.

Even with these advantages, video in VR is still video, produced to be screened using a TV, mobile device, or in a traditional movie theater and formatted to fit a rectangular frame. 360-degree video is a type of video that is native to the VR experience. It is a form of video without a frame that surrounds a viewer with video no matter where they look. 360-degree video is made using a special camera that captures a setting in all directions, including above, below, and behind the camera. Both fiction and documentary content

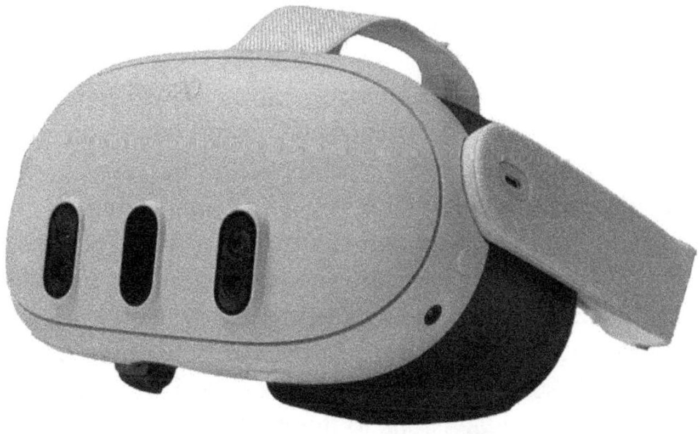

Figure 0.3. Meta Quest 3 VR Headset. *Photo by author.*

in immersive video places viewers in the center of the action, watching the story as it unfolds around them. As different as this is from traditional video, 360-degree video is still largely a passive experience: viewers cannot interact with characters in the story as it plays out in linear time. Like film and video, immersive videos have a beginning and end, and directors guide the story by choosing what to show and not show.

VR as a stand-alone experience uses our familiarity with television and movies to shape how we understand immersive stories. Other VR experiences are more interactive, from the VR rock climbing app The Climb to Superhot VR, a first-person shooter game. In these experiences, users interact with their environment using handheld controllers. The buttons and joysticks on the controllers resemble those for traditional video game consoles, and these experiences are related to games played on gaming systems. In fact, many VR games are adapted from traditional games and movie franchises, like Resident Evil 4 VR, The Walking Dead: Saints and Sinners, and Star Wars: Vader Immortal. Like the relationship between 360-degree video and traditionally framed film and television programs, VR games are differentiated from video games by losing the frame. This is easier to do than one may expect, as video games are often created in gaming engines that generate a virtual 3D environment. While a traditional video game allows players to change their view by moving a joystick controller, the VR headset itself is the controller in a VR game, producing a more immersive simulation. Video games provide a window into the world of the game, but VR games place players in the middle of the game.

Our experiences with video games, films, and online videos give us entry points into single-person experiences in VR. Multiplayer VR also has many precursors: social media apps, multiplayer gaming, group chats, and blogs all involve communication with other real people, either asynchronously or in real time, through a digital interface of a screen, keyboard, audio, or video. Live chats in games utilizing Xbox Live allow video game players to talk with one another while watching the characters they control on screen as they play. Mobile games like Roblox and games on Minecraft servers have become virtual teen hangouts, as friends play games while conversing. Zoom sessions and FaceTime group calls, especially after the onset of the COVID-19 pandemic, have become familiar ways of communicating.

All of this is to say that, even though VR is a novel experience, it is not unrelated to more established ones. People have had many experiences with precursor media technologies, and they bring those experiences as reference points for interacting in VR. Likewise, developers building VR apps often begin by applying these past examples to immersive experiences. When manufacturers first marketed televisions to the public, they described the invention as "radio with pictures."[18] This is what Marshall McLuhan described as "rear-view mirror" thinking: not understanding what is possible with a new technology or what effects it will bring, people tend to apply their knowledge of older, more established technologies to it.[19] Therefore, the early push for consumer VR relies heavily on concepts from gaming, online video, and social media. Is VR simply immersive gaming and video, or will it become something else?

Quickly, though, people begin to learn what a new technology is good for, and what it is not. Early VR educational experiences that recreate virtual classrooms filled with avatars of students sitting in auditoriums or desks aligned in rows mimic nineteenth-century models of education. While the experience creates a simulation of the real world, it is by no means an exact replica. We already have had similar experiences with remote schooling during the COVID-19 pandemic. Teachers who attempted to recreate the classroom experience on Zoom quickly realized its shortcomings, such as connection problems, screen fatigue, and the interrupting intrusions that became known as "Zoom bombing." Social inequities were highlighted, as students were seen in their home environments, and the quality of their connection depended on the technology they could afford. Even so, a new digital etiquette had quickly emerged, with phrases like "please have your cameras on" and "please mute your microphones" becoming commonplace. Like the new models that show how to better approach education in remote settings, new ways of interacting in VR are likely to arise. The ethical questions that the use of older technologies has generated may also apply to VR, while new ones are likely to emerge as VR is employed for new uses.

Related to this point, new technologies do not come out of nowhere; they rely on a foundation of related technologies already developed that make the new technology possible. VR in its current form could not exist without compact rechargeable batteries, small LED flat screens, computer and graphics processors, high-speed wireless data transfer, and the Internet. The same software packages that produce first-person perspective video games also are used to produce VR experiences. All these precursor technologies make VR possible, support its use, and shape its development.

In addition, the creation of new technology from old ones sometimes takes an evolutionary leap, generating new possibilities for doing things that make the technology even more desirable. Developers often seek to build "killer apps," application software programs that create new markets and demonstrate the functionality of new technology. The sitcom was the killer app for television, the spreadsheet and word processor for desktop computers, Super Mario Bros. for Nintendo video game consoles, and social media for smartphones. New technologies often experience a utopian period at first, with glowing promises of how that technology will solve multiple problems. Both radio and television were first promoted as educational tools with the potential for broadcasting lectures across the country, and Vice President Al Gore called the Internet an information superhighway and advocated connecting every school to it to access the world's knowledge.[20] Occasionally, a new technology, like the automobile, television, or smartphone, completely reshapes society.

Soon after a technology's widespread adoption, though, the utopian phase is followed by a dystopian one. After its explosive growth in the 1950s, by 1961, FCC Chairperson Newton Minow called television a "vast wasteland."[21] As late as 2012, social media was being praised for fostering the Arab Spring, a series of uprisings in support of democratic reforms, but by 2021, Congress was grilling Facebook about its role in spreading election and health disinformation, violating users' privacy, and harming the mental health of teens.[22] Praise for the Internet later gave way to concerns about cyber bullying, phishing scams, ransomware, and cyberterrorism. Despite the expectations that each technology would solve the problems for which they were intended, they also introduced unintended consequences that introduced new problems.

Educator and media theorist Neil Postman called this our Faustian bargain with technology. With every new technology we adopt, we gain some things but lose some others. No technology is a singularly positive or negative development, though some bring more positive or negative changes than others.[23] Understanding the changes a new technology will bring cannot be completely predicted, which begs the question: If we do not know what effects VR will have, should we even try to set up guidelines for its ethical use?

The answer to this question is yes, for the following reasons: While no one can predict the ultimate effects of any new technology, one can examine how that technology is being implemented and promoted and who is exercising control over it. Given that new technologies often rely on precursor technologies, they often share some of the same ethical issues. In addition, each technology extends power in some way, whether it is the automobile and our ability to travel or television and our ability to see events around the world. We have enough experiences with previous technologies to observe the results when we do not critically evaluate them, like the role of the automobile in suburban sprawl or the anonymity of the Internet leading to fake news, scams, and cyberbullying. Establishing an ethics of VR sets up a framework to critically evaluate its uses and effects as they occur, as well as predict and head off possible negative outcomes before they occur. While not outright rejecting the new technology, an ethics of VR works to prevent people from being seduced by it. Moreover, because VR's potential rests in its ability to not only simulate real environments but create entirely imagined ones, an ethics of VR asks about who has the power to imagine these worlds and share them with others and what virtual worlds people will be able to access. Once users enter a virtual world, how will they know if they are interacting with other real people or bots, and whether data about where people have visited and what they have done there will be stored and used. An ethics of VR helps to set some ground rules for the metaverse and encourages users to think about the effects the technology will have on them and others.

THE CHAPTERS IN BRIEF

Before beginning a discussion about the ethics of VR, it is important to understand first what VR is. Chapter 1 discusses how VR works and differentiates it from similar experiences like AR, XR, and mixed reality (MR). It explores the history of VR's development and examines how it takes advantage of human perception, concluding with predictions about the future of VR and the possible ethical issues that future will generate.

The second chapter reviews the study of ethics, from its foundations in philosophy to its application to technology and media. A system of ethics is more than proclaiming the "right" thing to do in each circumstance. There are systematic ways of thinking through issues that consider various stakeholders and the interplay between decisions that favor one group while placing burdens on others. Should every ethical decision benefit the most amount of people, or are there some possibilities that should never be considered? An ethics of VR seeks to delineate these choices and evaluate them in a manner that goes beyond what may simply feel like the right decision. The ethics laid out in this book is a form of applied ethics, a consideration of guidelines that are practical in nature to address the social and political issues that emerge when a new technology is introduced that changes the relationship of senses and communication. The ethical discussions in this book are both descriptive: what situations may occur that require making an ethical decision, and prescriptive: what one *should* do in a situation.

After describing what VR is and examining how the field of ethics can be applied to navigate this new technology, the following chapters outline groups of ethical issues that either already exist or will likely soon become topics of concern for VR. Examining related media technologies, like social media and AI, these chapters discuss ethical concepts already applied to older media technologies and how they can be used to think about VR, as well as how older concepts may not be enough to deal with the unique attributes of VR and other related immersive media. The third chapter examines the ethics of control over VR hardware, applications, and users. With Meta controlling the vast majority of installed VR headsets in the market, what outsized power do they have in shaping the future of the metaverse? How much control should developers exercise over the virtual spaces they design, and how much control should be ceded to users? Moderators have served a useful role in older media that host virtual communities, but who should be held responsible when there are conflicts or accusations of abuse within moderated spaces? Finally, how does the technology itself shape and limit the activities that take place in VR? Different ethical frameworks, including utilitarianism, human rights, and duty of care ethics, may have different approaches to these questions.

Chapter 4 outlines the challenges to accessing VR environments. Beyond the price of a VR headset, the cost to participate in immersive environments often includes a high-speed Internet connection and associated computer hardware. The digital divide has been an active issue since the 1990s, when the Internet became available to the public and schools, governments, and companies began to offer services online. The divide is amplified by access to more basic services, like consistent electricity and shelter. Even when these challenges are overcome, many people may not be able to effectively utilize a medium that relies on sensory immersion and mobility. What obligations do governments, private industry, and individuals have in ensuring access to VR as more opportunities and services move into the metaverse?

The fifth chapter, "Privacy in VR," explores how older digital technologies made the collection and dissemination of private data a big business, and the capacity for VR to provide even more intimate data to marketers and state actors. With the ability to measure pulse, pupil dilation, eye gaze, patterns of movement, and infer emotional states, VR headsets offer more insights about users as they enjoy participating in immersive environments. As these data points are processed by AI and other algorithmic systems, users are at risk of surveillance and data discrimination. While some data sharing is required for multiuser immersive spaces to function, the process for users to agree to share their data is often opaque and favors developers and data brokers over individuals. The power imbalance between users and the large companies and governments that make use of their personal data is a condition that ethical limits can address.

Chapter 6 shows the capability for VR to represent environments in misleading ways to manipulate users, and how developers and users bring into the medium the implicit biases and stereotypes that they have internalized from their social environments. While VR users can adopt avatars of their choosing, verification of accurate representation may become a norm in the future for some virtual spaces. A form of digital blackface, in which users of a dominant group put on avatars representing traditionally oppressed groups, does the opposite of VR pioneer Chris Milk's vision of VR as an empathy machine. When the designers of a system belong mostly to a single group, they tend to unconsciously build that group's biases into that system, reproducing the same social inequities in the virtual world that exist in the physical one. Some VR apps, though, challenge the biases in the system by foregrounding them and calling out the characteristics of a supposedly neutral technology, educating users on their own implicit biases.

The seventh chapter examines cases in which participants in VR have been harassed by other users and imagines future cases where abuse can be inflicted upon others. Digital harassment did not begin with VR: the infamous Gamergate episode revealed the rampant misogyny lurking in video game communities, as well as the positive responses to address harassment. Immersive environments have the potential for even greater levels of abuse. The perception of space and proximity to other avatars creates a feeling of vulnerability, especially when those spaces are under the control of others. Without clear ethics of control and privacy limiting malevolent actors, the possibility to abuse others in the medium carries real impact. While VR users can remove their headset and return to the physical world, the increasing blending of the virtual and physical worlds may make it more difficult to opt out of environments where harassment and abuse take place. Building civil spaces that respect the autonomy of each individual should be an ethical goal for VR.

Finally, the last chapter explores what an applied ethics of VR can suggest for imagining future uses and limits of the technology. Given the possibilities for corporate or state control, violations of privacy, misrepresentation, and abuse, the metaverse can become the dystopian environment imagined by science fiction. However, careful consideration of the possibilities and misuses of the technology can guide its development to become a useful and liberating medium that empowers individuals while allowing them the freedom to explore, imagine, and build new possible environments that may hold solutions to problems in the physical world.

The purpose of asking these questions about VR is not to prevent these technologies from being developed or used. Oftentimes, promoters of new technologies dismiss critics of those technologies as Luddites, referring to the loom-smashers of the nineteenth century in England who smashed the mechanical looms that were putting weavers out of work. Critics are accused of being afraid of new technologies or members of an old guard trying to preserve a *status quo*. However, critically analyzing new technologies before they are adopted in a widespread manner promises the possibility of a degree of control over them, rather than allowing the biases of a technology to sweep over people before they have a chance to adapt or defend their own sense of humanity. Asking these questions now may reveal opportunities to develop systems of maintaining a balance so that people are not beholden to the technology or the few people who have identified how to use the technology at an advantage over others. Ultimately, we should ask these questions so that we may adopt VR to serve us well, instead of becoming subservient to it and the companies that control it.

I would be remiss in not acknowledging that the company Meta (formerly known as Facebook) is mentioned many times in this book. This is because Meta is a large player in the development and promotion of VR. However, it is not the only player, and history demonstrates that no one company controls a technology forever. Other major players, including Apple and Google, have built VR headsets and software to create immersive environments. The metaverse, like the Internet, likely will welcome virtual developers from around the globe. In fact, Meta has announced that it is permitting other headset manufacturers to run its Horizon OS software, allowing interoperability across devices.[24] In the end, even though several cases presented here examine the ethical stances and consequences of Meta, this is not a book about Meta or its practices. Rather, these examples illustrate the questions and problems we will face as VR becomes a more dominant technology that interweaves itself into more of our daily lives.

By the way, the imagined example presented at the beginning of this chapter is not so far away. The company VictoryXR partnered with 10 colleges in 2022 to launch "metaversities," a virtual space where students attend classes in VR while using Meta Quest headsets. With the ability to manipulate virtual 3D objects, students can access their classes from anywhere with an internet connection and see digital reproductions of college campuses.[25] Critics already are expressing privacy concerns over Meta's collection of student data, and Apple has announced that it will not collect data about where users are looking while wearing the Vision Pro headset, though it uses a retinal scan of users' eyes to unlock the device.[26] In a survey published in an article in *The Chronicle of Higher Education* about this development, only 22% of students said they understand how their institution uses their personal data, and only 49% trust that their institution uses their private data ethically and responsibly.[27] This shows why it is important to examine ethical issues about VR before it is widely implemented.

NOTES

1. The marketing of this app as an intense immersive experience is emphasized on its website: https://www.getsupernatural.com.

2. The VR game engine Unity has published several articles about the usefulness of VR in these areas: Gajsek, Dejan. "Virtual Reality Therapy as Exposure Therapy Alternative: Early Studies, Best Examples, and Possible Future of VR Therapy for Mental Health." *3D Development Bootcamp & XR Courses*, June 1, 2022. https://circuitstream.com/blog/virtual-reality-therapy/.

3. Apple, Inc. "Introducing Apple Vision Pro: Apple's First Spatial Computer." *Apple Newsroom*. Accessed April 12, 2024. https://www.apple.com/newsroom/2023/06/introducing-apple-vision-pro/.

4. You can explore these VR applications by following these links: Tribe XR: https://www.tribexr.com/; SynthVR: https://store.steampowered.com/app/1517890/SynthVR/; Tilt Brush: https://www.tiltbrush.com/; Gravity Sketch: https://www.gravitysketch.com/; "The Night Café": https://www.oculus.com/experiences/gear-vr/908477449239883/; "Traveling While Black": https://www.oculus.com/experiences/gear-vr/1994117610669719/.

5. Gartenberg, Chaim. "Meta's Oculus Quest 2 Has Shipped 10 Million Units, According to Qualcomm." *The Verge*, November 16, 2021. https://www.theverge.com/2021/11/16/22785469/meta-oculus-quest-2-10-million-units-sold-qualcomm-xr2; "Virtual Reality Market Size Worth $87.0 Billion By 2030." *Grand View Research*, April 2022. https://www.grandviewresearch.com/press-release/global-virtual-reality-vr-market.

6. Wachowski, Lana, and Lilly Wachowski, dirs. *The Matrix*. United States: Warner Bros, 1999. Users were able to download the Unreal Engine 3D environment of *The Matrix* here: https://www.unrealengine.com/en-US/blog/introducing-the-matrix-awakens-an-unreal-engine-5-experience.

7. Kuperczko, Diana, Peter Kenyeres, Gergely Darnai, Norbert Kovacs, and Jozsef Janszky. "Sudden Gamer Death: Non-Violent Death Cases Linked to Playing Video Games." *BMC Psychiatry* 22, no. 1 (December 23, 2022): 824. https://doi.org/10.1186/s12888-022-04373-5.

8. A restored version of the Lumière short can be seen here: *[Actual 4K Scan] The Arrival of a Train at La Ciotat Station—Lumière Brothers—1896*, 2020. https://www.youtube.com/watch?v=1FAj9fJQRZA.

9. Cline, Ernest. *Ready Player One*. 1st ed. New York: Crown Publishers, 2011; Spielberg, Steven, dir. *Ready Player One*. United States: Warner Bros, 2018.

10. Stephenson, Neal. *Snow Crash*. Bantam Spectra Book. New York: Bantam Books, 2008.

11. Bidar, Musadiq, and Dan Patterson. "Facebook Bets Big on the Metaverse. What Is It?" *CBS News*, November 1, 2021. https://www.cbsnews.com/news/metaverse-facebook-virtual-augmented-reality-hubs/.

12. Robertson, Adi. "Facebook Is Making Oculus' Worst Feature Unavoidable." *The Verge*, August 19, 2020. https://www.theverge.com/2020/8/19/21375118/oculus-facebook-account-login-data-privacy-controversy-developers-competition.

13. Moyer, Edward. "Apple AR, VR Headset to Feature Content from Hollywood Directors, Report Says." *CNET*, June 4, 2022. https://www.cnet.com/tech/computing/apple-ar-vr-headset-to-feature-content-from-hollywood-directors-report-says/.

14. Microsoft. "Microsoft HoloLens: Mixed Reality Technology for Business." Accessed September 24, 2023. https://www.microsoft.com/en-us/hololens.

15. Cross, Jason. "Apple Vision Pro: What Is Spatial Computing Anyway?" *Macworld*, February 2, 2024. https://www.macworld.com/article/2215989/apple-vision-pro-what-is-spatial-computing.html.

16. A good collection of holodeck episodes can be found here: Weyer, Michael. "*Star Trek*: Ranking the 20 Best Holodeck Episodes." *CBR*, January 4, 2019. https://www.cbr.com/star-trek-best-holodeck-episodes-ranked/.

17. Harris, Owen, dir. "San Junipero." *Black Mirror*. Netflix, October 21, 2016. https://www.netflix.com/watch/80104625.

18. Van Den Ende, J., W. Ravesteijn, and D. De Wit. "Shaping the Early Development of Television." *IEEE Technology and Society Magazine* 16, no. 4 (1997): 13–26. https://doi.org/10.1109/44.642518.

19. McLuhan, Marshall. *Understanding Media: The Extensions of Man*. 1st MIT Press ed. Cambridge, MA: MIT Press, 1994.

20. Gore, Al. "Inauguration of the First World Telecommunication Development Conference (WTDC-94)." *Speech, Buenos Aires*, March 21, 1994. http://search.itu.int/history/HistoryDigitalCollectionDocLibrary/4.144.57.en.104.pdf.

21. Minow, Newton. "American Rhetoric: Newton Minow—Address to the National Association of Broadcasters (Television and the Public Interest)." Accessed September 24, 2023. https://www.americanrhetoric.com/speeches/newtonminow.htm.

22. Kang, Cecilia. "Facebook Grilled by Senators Over Its Effect on Children." *The New York Times*, September 30, 2021. https://www.nytimes.com/2021/09/30/technology/facebook-senate-hearing.html.

23. Postman, Neil. *Technopoly: The Surrender of Culture to Technology*. New York: Vintage Books, 1993.

24. Barr, Kyle. "Meta Quest Horizon OS Will Finally Make 3rd Party VR Headsets Worth Buying." *Gizmodo*, April 22, 2024. https://gizmodo.com/meta-quest-horizon-vr-quest-3xbox-1851426974.

25. Swaak, Taylor. "What Colleges' Contracts for 'Metaversities' Leave Out." *The Chronicle of Higher Education*, June 17, 2022. https://www.chronicle.com/article/will-virtual-reality-metaversities-risk-students-data-privacy-we-asked-for-contracts-to-find-out.

26. AppleInsider. "Learn about Apple Vision Pro," June 8, 2023. https://appleinsider.com/inside/apple-vision-pro.

27. Swaak, "What Colleges' Contracts for 'Metaversities' Leave Out."

Chapter 1

What Is VR?

Virtual reality (VR), XR, AR, mixed reality, 360-degree video, Computer Automatic Virtual Environments (CAVEs), head-mounted displays (HMDs), avatars, haptics, and positional audio: with any new technology, and digital media technologies in particular, the amount of jargon and acronyms can become dizzying. As with any new technology, a core group of early adopters and developers create new names to differentiate their products and functions. With the use of these terms, sometimes it is difficult to separate the hype from reality. Before an ethics of VR can be considered, it is important to first understand what VR is. This technology, like any complex digital system, is made up of components that each has a history of development. It is important to understand the differences between VR and older media in order to determine what VR is good for and why it has taken on the functions that it has. In particular, VR engages the human senses and takes advantage of their idiosyncrasies to create the illusion of immersion. With this understanding and by examining recent innovations in VR, it is easier to imagine the potential futures of this immersive medium.

HOW VR WORKS

As it exists today, VR usually is achieved with a HMD that blocks out the view of the real world and replaces it with recorded video or computer-generated moving images. Each eye is fed separate offset images, creating a stereoscopic image that creates an illusion of depth, commonly referred to as 3D. As a user wearing the HMD moves their head, the point of view of these images changes in real time, creating the illusion that the user has been transported to the center of a virtual space. Whereas traditional films and videos display images within a frame (a window to the world), VR surrounds the user with images, in front and behind, up and down. This is why VR is often referred to as immersive media.

In addition to immersive video, most HMDs have built-in headphones that feed positional audio to the user. As the user moves their head, the perceived direction of audio changes along with that movement, so that a sound seemingly coming from a specific direction will remain located there. If the user moves their head to the side, the audio will sound as if it is coming from that same direction. Some VR experiences use positional audio to guide users to look in a specific direction. Hearing a sound coming from behind them will cue people to turn around and look at the direction from where the sound seems to be originating.

Some HMDs are equipped with eye trackers to record where users are looking within their field of vision. These eye trackers allow people to use eye movements to interact with the immersive environment, including having virtual characters notice when a user is looking at them. Eye tracking also allows for foveated rendering, a more efficient method of display that provides a higher-resolution image at the spot where a user is looking. Eye trackers can also record a user's pupil dilation and pulse, which provide feedback that tracks physical activity and can infer the emotional state a user is experiencing.[1]

To help interact in the virtual environment, most HMDs are accompanied by hand controllers that include a joystick, several buttons, and triggers. These are more than a simple controller, however, as they send positional data to the VR software, enabling virtual representations of a user's hands to appear in the HMD. Some HMDs now use cameras to track a user's hands so that a controller is not necessary. Instead, users can make hand gestures, like swiping or pinching, to execute commands. Meta Quest 3, for example, can run software that allows users to switch between controllers and hands. (See figure 1.1.)

While not nearly as widespread as HMDs themselves, VR accessories can heighten the immersive experience. Haptic vests, suits, and gloves provide feedback like vibrations, pressure, and low-frequency sounds to different parts of the body. Special VR 360-degree treadmills or sliding disks allow users to move around in VR more naturally. Instead of using a joystick, users literally walk or run on these stationary platforms to move in the VR environment. Steering wheels, flight controllers, or guns can replace handheld controllers to simulate interacting in specific VR environments. Facial trackers record facial expressions and confirm a user's identity, and room-based motion trackers can improve location placement in the virtual environment.

While VR content is best viewed through HMDs, less expensive enclosures can produce a similar experience. Google initiated its Google Cardboard

Figure 1.1. Left: Using a Meta Quest VR HMD. Right: What Is Seen in the HMD. *Photo and screenshot by the author.*

Figure 1.2. Left: Google Cardboard. Right: An Opened VR Smartphone Enclosure. *Photo by author.*

project in 2014, which utilizes a user's smartphone for the display. For a nominal fee, Google shipped to people who requested it a flat piece of cardboard and two plastic lenses that could be folded into a viewer.[2] Users can insert their phone and hold the viewer up to their eyes to view a VR experience. The phone displays two side-by-side images, one for the left eye and the other for the right, which produces the stereoscopic 3D image. People also can download a template for free to cut out and self-assemble their own cardboard viewer or have one printed on a 3D printer. In 2015, *The New York Times* shipped to its subscribers a *Times*-branded cardboard viewer and encouraged people to download its new VR app.[3] Since these early innovations, manufacturers have produced various plastic enclosures with head straps that use smartphones in a similar way. (See figure 1.2.)

A third way to view VR content is by simply using the smartphone itself. By moving the phone in different directions, users can see the entire virtual space on the screen. Users can interact with the VR experience by tapping directly on the touchscreen of the phone. However, this does not create nearly as immersive an experience, as users view the space through the window of the phone's display, much like the frame of a television or film screen. The image is strictly two-dimensional, and sound originates from the phone, negating the possibility of positional audio. Viewing VR content this way is a bit like looking through a submarine's periscope: the virtual space can be seen in all directions, but only through the narrow view of the frame.

An alternative, though far less portable, way to experience VR is by using a CAVE, a room consisting of three walls and a floor onto which images of the virtual world are projected.[4] A benefit of CAVEs is that they can accommodate multiple users who all can stand in the room, and it cause less VR-induced motion sickness than HMDs sometimes do. Because of their size and cost, CAVEs are usually found only in labs and large companies and are used for simulations, design, and research. Some CAVEs use 3D technology that requires the use of glasses to perceive the stereoscopic projection.

Experiencing VR with an HMD is a unique experience that cannot be fully described in words. Under the best conditions, it can generate an illusion of presence that metaphorically transports users to another place. However, there are still lingering issues that developers are working to solve. The first is the field of view. Humans typically have a binocular horizontal field of view of 200 degrees and a vertical field of view of about 135 degrees. The most popular HMD, the Meta Quest 3, can reproduce only about 110 degrees horizontal and 96 degrees vertical field of view. In addition, eyes saccade (move rapidly) three times per second on average as people visually scan their environment. A limited field of view results in less immersion, as peripheral vision is not supplied with an image. However, this problem likely will be solved soon: already, the Pimax Vision 8K, the most advanced commercially available HMD, can produce a horizontal field of view that approaches 200 degrees, and Apple's Vision Pro boasts 23 million pixels that produce a 4K image for each eye.[5]

A second problem with HMDs is their resolution as measured in pixel density. Pixels (picture elements) are dots that each combine different intensities of red, green, and blue primary colors of light and are arranged in rows to produce a video image. If viewers are far enough away from a screen, they see an image, but as they move closer, they can begin to resolve individual pixels. A person can experience this by viewing a jumbotron up close or looking at a television screen through a magnifying glass; instead of an image, they will see individual glowing red, green, and blue vertical bars.

Immersion is helped when the pixels in an image are close enough together that users see a complete image instead of individual pixels. Most people cannot resolve a pixel density greater than 60 pixels per one degree of view. However, many HMDs, including the Meta Quest 3, have a density only of about 25 pixels per degree. This results in what VR experiencers call a "screen door effect," or the perception of a black grid over the virtual environment as if they were viewing it through a screen door. Like the field of view, the pixel density problem also is likely to be solved as higher-resolution screens with smaller pixel pitches are developed. Both the Meta Quest Pro and Apple Vision Pro have reduced this effect with a greater pixel density that surpasses the screen door threshold.[6]

Another issue that most HMDs have solved recently is their refresh rate. People typically see a sequence of still images as still images up to about 12 frames per second (fps), but they can perceive individual changes to a frame up to about 77 fps, or 13 milliseconds. Cinema uses a rate of 24 fps to create the illusion of motion, but the flicker and strobing effects it causes are what people have come to associate as the premium experience of "film." By the way, this is why movies shot on video use this lower frame rate, even though most cameras can shoot at higher rates. Most HMDs use a refresh rate of 90 or 120 fps, but many VR games designed for them can produce only 60 fps. As game development software improves, it will catch up to the hardware and produce more lifelike moving images.[7]

Motion tracking and latency also affect immersion. These involve how quickly the visual image is updated in the HMD in response to head movements. Motion tracking tells the VR software of the changing position and orientation of the HMD so that the software can update the display to the new field of view. In the real world, people's eyes receive light traveling at almost 300,000 km/s that is reflected from objects; when they turn their heads to a new direction, they see reflected light from that direction. When users turn their heads in a virtual world, the motion tracking of the HMD sends updated positioning to the VR software, which must compute and send the new view to the display. The delay in processing and transferring frames of video is called latency. Like the buffering of a Zoom call interrupting a conversation, latency breaks the illusion of immersion. It also is a major contributor to motion sickness, a common problem with VR. It is impossible to eliminate latency completely, but a delay of 15 milliseconds or less seems to be acceptable for most users. Many factors contribute to latency, including refresh rate, data transfer rate, processor and memory speed, graphics rendering hardware, and the VR software itself.[8]

Even when using the same HMD, users can have many different types of VR experiences. Recorded or live streamed VR video, also called 360-degree video, shows photorealistic video and audio in all directions. This type of experience is described as having three degrees of freedom (3DoF), where users can change their orientation (pitch, roll, yaw) but not their position. This video typically is captured with a 360-degree camera, like the GoPro Max or Ricoh Theta. These cameras have two fisheye lenses that capture action in front, behind, above, and below the camera. Some cameras, like the Vuze+ 3D or the Insta360 Pro, use multiple lenses to record several streams of 3D video that are stitched back together to create the VR experience.[9] Even though watching 360-degree video is much more immersive than framed video, it is otherwise a passive experience. Videos have a beginning, middle, and end, and users cannot interact with the video other than cueing and varying the playback speed of the video. An advantage to 360-degree video is that it is relatively easy to produce, captures photorealistic imagery, and can be watched sitting or in an otherwise stationary position. While it is more immersive than framed video, many users feel as if they are watching an event through someone else's eyes rather than accepting the illusion that they are themselves in the space.

Interactive VR experiences require another form of production altogether. Instead of capturing content with cameras, these experiences are created in 3D modeling, animation, and gaming software. Gaming engines like Unity, popular for producing many mobile games, and Unreal, used to produce the wildly successful multiplayer game Fortnite, are designed to render 3D environments that players can move through in real time. For VR experiences, the HMD takes the place of a game controller to move the player through the environment, and hand gestures or controllers allow users to interact with characters and objects. Gaming engines are particularly useful for VR because they were designed to solve the problem of updating images based on a player's position and orientation. Beginning with the first-person shooter game Doom in 1993, video games have evolved to allow for real-time interaction between multiple players. For example, Fortnite allows for up to 100 players in the same match, with each player seeing their own first-person view of the game.[10] (See figure 1.3.)

Because of the processing power needed to generate different views in real time, most interactive VR experiences appear as animations or somewhat

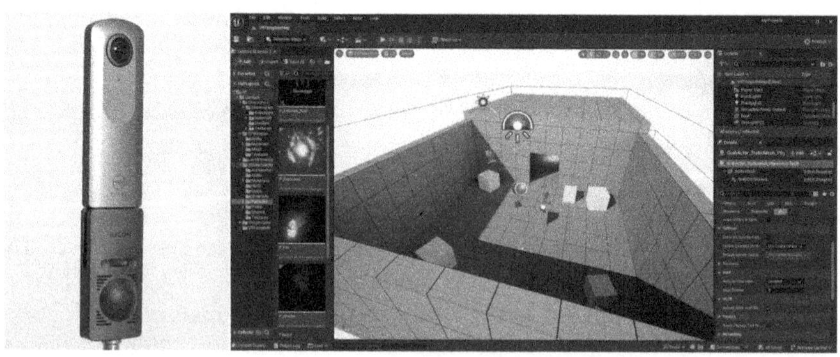

Figure 1.3. Left: Ricoh Theta V 360-Degree Camera. Right: The Unreal Engine Environment. *Photo and screenshot by the author.*

unrealistic representations. Like during the early days of video games, VR developers are constantly battling memory and storage limits for games of increasing complexity. Though this problem will be solved with increased computer capacity, a danger with attempting photorealism in VR is almost achieving it. When animated images are close to realistic but not quite perfect, like the characters in the children's film *The Polar Express*, observers experience an unsettling feeling that has been labeled the uncanny valley.[11] The animation company Pixar has reduced the photorealism of several of its films' characters by stylizing them or animating non-human forms, like toys, cars, and robots, to avoid the uncanny valley. However, the release of Unreal Engine 5 allows for more photorealistic rendering that will continue to improve the look of interactive VR experiences, and programs like Unreal's MetaHuman software promise to generate animations that are indistinguishable from real people. Unreal Engine already is used in virtual productions for film and television, like *The Mandalorian*, where computer-generated backgrounds are seamlessly composited with live actors in real time. Advances in image-producing AI programs have approached photorealism, though sometimes they produce surreal images and videos that are photorealistic fantasies, like Darth Vader holding a fish or a dragon swimming in pudding.[12]

While some interactive VR experiences, like 360-degree video, are limited to 3DoF, many interactive experiences provide full six degrees of freedom (6DoF), allowing for not only changes in orientation but also movement side to side, up and down, and back and forth (often mapped with an X-Y-Z axis). One danger of 6DoF for users is running out of physical room to navigate the virtual world. The Meta Quest solves this problem with its Guardian system, where it instructs users to draw out a boundary of safe space cleared for 6DoF.[13] When a player approaches the boundary, a grid appears, and the Quest shows a view of the real world from its onboard cameras. Even so, users have experienced many cases of tripping over furniture, bumping into walls, or smacking a controller-wielding hand as they swing it to attack a virtual enemy. One developer told me a story about someone trying out the VR game Walk the Plank: the player saw in their HMD that they were walking on a plank extending from a ship and suspended over the ocean. Suddenly, the player decided to dive off the plank and into the water. Of course, instead of landing in water, the person smashed into the floor and required first aid.

When designing ways to move in VR, developers must think about two spaces simultaneously: the virtual space of the VR environment and the physical space of the VR user. Given the problem of limited physical space for experiencing 6DoF VR, programmers have developed several ways of moving through virtual worlds. One way is borrowed from cinema: when changing from one space to another, the experience fades out of the old space and into the new one. While audiences have accepted this as a convention in cinema to transition between scenes, in VR the fade out and in sometimes create the illusion of losing and regaining consciousness. Another method is called teleporting, where users are instantly transported from one virtual space to another by selecting the space where they want to be. To prevent motion sickness from sudden position changes, developers also have tried visually logical methods of locomotion, like vehicles. Users who want to move to a new space hop into a virtual car or plane which brings them to their destination. As it is closely related to their experience of sitting stationary in a moving vehicle, this reduces the chance of motion sickness.[14]

When people put on an HMD, they can have many different types of VR experiences. One of the most popular is gaming. Many popular video game franchises have produced VR versions, including Fallout, Half-Life, and Resident Evil, as well as popular strategy and action games like Beat Saber, Fruit Ninja, and the time-manipulation game SuperHot. Simulators are also popular, like racing and flight simulators, the free climbing game The Climb, and even satirical simulations like Job Simulator and Vacation Simulator. Beyond gaming, one of the most successful VR apps is Supernatural, a subscription fitness app featuring coaches in otherworldly settings. Users can paint and sculpt in virtual space with Tilt Brush and Gravity Sketch, and they can attend concerts and performances remotely in virtual spaces. Businesses and educational institutions have begun to use VR for training simulations, especially for scenarios that are too dangerous or expensive to attempt in real life. Doctors now use VR to plan out surgeries by immersing themselves in virtual models of their patients built from MRI and CT scans. People can virtually travel to anywhere in the world, or out of this world, from the bottom of the ocean to the International Space Station. Remote collaboration and project management software allow teams to meet virtually and may help solve the remote work problem by offering real-time collaboration without Zoom fatigue. Psychologists have been using VR for exposure therapy; for example, the VR app Spiders! helps patients suffering from arachnophobia. Meta is betting big on social VR, using its social networks to build what resembles, to some, an immersive version of Second Life. During the launch of its Quest 2 HMD, it ran a commercial in which a girl and her grandfather, living in two different locations, are fishing together in a virtual stream. As a company working to make VR a central part of the metaverse, it is building virtual spaces where people meet to work, learn, and play.

AR, XR, AND MR

The technology that makes VR possible can also be employed for related experiences. Instead of completely replacing the real environment with a virtual one, AR involves displaying text, graphics, images, or animations over a view of the physical world. Developers plan to use AR to provide contextual information when viewing people, places, and objects, such as a stranger's Facebook or LinkedIn profile or information about a painting at a museum. One advantage of AR is that it can be integrated with a smartphone and camera, equipment many people already possess. AR has received support from advertisers, who hope to display location-based offers as people view the world through AR-equipped devices like smart sunglasses.

Companies are working to integrate AR into a lightweight heads-up display (HUD). Google experimented with building a wearable HUD with Google Glass, eyeglasses that displayed data projected on the lenses, beginning in 2013. Its Google Glass Enterprise 2 edition, like Microsoft's HoloLens, is marketed to businesses to be used for accessing technical information in a hands-free way. Meta partnered with Ray-Ban in 2021 to produce Ray-Ban Stories Smart Glasses, stylish sunglasses that have an embedded camera, microphone, and miniature speakers. Apple's Vision Pro displays Apple's familiar iOS app icons in a user's field of view, and a dial on top of the headset allows users to adjust the amount of immersion they experience. As the technology improves, some predict that these HUDs will replace smartphones as a primary interface.[15]

If AR provides digital information about the physical environment, MR goes further by mixing interactive virtual content with the physical world. MR requires scanning the physical environment so that the virtual content can interact with it. Many people had experienced an earlier version of MR when they downloaded and played the mobile game Pokémon Go. In 2016, Niantic released the location-based game, in which animated characters appeared over real-world locations when viewed through the phone's camera. While many observers at the time called this game an AR app, Pokémon Go is more accurately labeled MR because the animated content responds to users and the environment rather than merely providing more information about a physical space. Instagram, TikTok, and Snapchat filters all employ MR technology, mapping out the camera's view and overlaying graphics that travel with objects and faces in the frame.

A term that encompasses VR, AR, and MR is extended reality (XR), which refers to any technology that combines the virtual and real worlds, up to and including completely replacing the real with the virtual. By extending and engaging human perception with digital tools, people's senses can perceive data, images, and sounds far beyond what they could do naturally. Some HMDs that combine functionality between VR, AR, and MR are called XR headsets, and most future headsets will be XR, with the ability to switch between different modalities. From scientific visualization to information sharing to entertainment, XR promises to transform the human experience.

A BRIEF HISTORY OF VR

However new it may seem to be, the development of VR has a long history. VR is inspired by and is an extension of human storytelling itself. The prehistoric Chauvet cave paintings took advantage of the contours of the rocks to make their images appear three-dimensional. After the invention of photography, the stereoscope allowed nineteenth-century Victorians to see images in 3D, and panoramic paintings that depicted battlefields and far-away landmarks were installed in rotundas and provided immersive experiences for guests.

As soon as film and video screens could display moving images, inventors attempted to find ways to use them to immerse the senses. The cinematographer Morton Heilig patented a machine in 1962 that he called Sensorama. Sitting at a large console, a user would place their head in a hood that blocked outside light, see stereoscopic color monitors, and hear stereo sound. Tiny fans would blow air and odors to recreate the smells of a scene, like a motorcycle buzzing through the streets of New York City or a helicopter ride. While it never became commercially viable, the Sensorama was one of the first systems that attempted to engage multiple senses to create a sense of immersion.[16]

Two years earlier, Heilig had patented in 1960 the Telesphere Mask, credited as the first HMD. The Mask used two small CRTs (cathode ray tubes, or traditional video monitors) to generate 3D video, but it did not have the capability to update the field of view as users turned their heads. That ability came eight years later with the Sword of Damocles. Ivan Sutherland, a computer scientist who had worked at IBM and Harvard before he moved to the University of Utah in 1968, focused his work on computer graphics and digital user interfaces. The Sword of Damocles was named after its appearance, with giant rods suspended from the ceiling and attached below to what Sutherland called the Ultimate Display. As the wearer of the display would turn their head, leads attached to the rods would provide feedback to a computer, which would update accordingly a simple wireframe drawing displayed in the HMD. This invention is credited as the first time that object stationarity was demonstrated in VR, or the perception of seeing an object remain in a stationary position as an observer changed their point of view.[17]

While these early experiments demonstrated the possibilities of VR and immersive media, it was not until the 1980s that commercial headsets were produced. In 1987, VPL Research, co-founded by Jaron Lanier, produced the EyePhone, an HMD that displayed images at about six frames per second, for large corporate customers. The EyePhone worked with VPL's other inventions, the DataGlove and DataSuit, which provided interfaces for interacting with the virtual environment. Lanier, a former employee of the video game company Atari, is credited with coining the phrase "virtual reality." He envisioned a future when people could use these interfaces like musical instruments to improvise virtual environments.[18] The EyePhone was a major advancement in VR and resembled what HMDs look like today, but it was far too expensive at the time to be financially feasible, and VPL Research declared bankruptcy in 1990.[19]

During the 1990s, virtual environments were imagined in movies, but advances in VR itself were limited to research labs, the military, and NASA. Several home video game console companies attempted to build VR interfaces, but none were successful. However, once digital processing, data speeds, and video graphics cards improved and flat-screen LED screens became feasible, the industry was ready to take another leap forward. In 2012, 19-year-old Palmer Luckey created a Kickstarter campaign to raise $250,000 to build an affordable HMD. He called his new company Oculus, and the product was named the Oculus Rift. The HMD used a seven-inch screen and 24-bit color to display side-by-side stereoscopic images. Luckey promised a developer's kit for a $300 donation, which included the headset and a copy of the video game Doom to play in VR. By the end of the campaign, almost $2.5 million had been raised from over nine thousand backers.[20]

The Oculus Rift ushered in the current era of VR experiences. Oculus shipped the original development kit in 2013, and the first commercial version available to the public arrived in 2016. By then, Facebook had acquired Oculus for $2 billion. It followed the Rift with Oculus Go, a stand-alone wireless HMD that did not have to be tethered to a PC to operate. Facebook launched an Oculus App Store, which functioned like the Apple App Store and Google Play, where users could purchase and download new VR apps. By 2022, its Quest 2 headset had sold more units than the gaming consoles Xbox X and S combined.[21] The company has since launched the Quest Pro, a headset priced just under $1,000 with improved resolution and a faster image refresh rate, and a lower-priced Quest 3. As a division of Meta, Oculus has since been renamed Reality Labs, and Meta has committed to a strategy of developing the metaverse as a virtual space for social interactions.

While Meta dominates the current market for VR HMDs, its success has inspired many competitors to build competing units. One of the most impressive is the HTC Vive. Though it costs more than twice as much as a Quest 2, the Vive Pro 2 has a 5K display, a 120Hz refresh rate, and 120-degree field of view. Sony has launched its own HMD with the PlayStation VR. Designed to work with the PlayStation gaming console, the headset uses an OLED screen to create 360-degree views of PlayStation VR-enabled games. The Valve Index is a headset designed to work with the Steam VR distribution platform, which provides gaming content to several brands of VR HMDs. The Pimax Vision 8KX DMAS boasts an 8K screen, the highest resolution of any headset, with promises of producing a 12K version soon. However, Meta's largest competitor by far is Apple, which already dominates the smartphone market with its iPhone. In 2024, Apple launched the Vision Pro, a headset with 23 million pixels, 12 cameras, 6 microphones, and 5 sensors.[22] Though its $3499 price tag puts it out of reach for most consumers, the device ushers Apple into what it calls spatial computing, an XR experience that adjusts from simple AR applications to fully immersive VR.

VR AND HUMAN PERCEPTION

All of these headsets use video and audio to create VR content, but people use more than their sight and hearing to make VR an immersive experience. To understand how human senses interact with VR, it is important to first realize that people have more than the five senses that they learned about in grade school. Beyond the senses of sight, hearing, smell, taste, and touch, people also have a vestibular sense, commonly known as the sense of balance. The semicircular canals in the inner ear provide feedback to the brain about a body's position relative to the ground, as well as linear and radial acceleration.

In addition to the vestibular sense, people also have the sense of proprioception, or the awareness of the position of their body in space. Proprioception provides feedback about the position of joints in the body, whether and how they are moving, and if they are meeting resistance from a force. Neurons located in the muscles and joints throughout the body send information to the brain so that it can make decisions about corrective motor actions. If the vestibular system helps a person understand whether they are balanced or are about to fall, proprioception helps to keep balance by correcting the body's position. Reaching for objects and voice modulation are also aided by proprioception. A person can become aware of proprioception by closing their eyes and moving their arms to different positions.

What is commonly thought of as the sense of touch, or the haptic sense, actually is made up of several distinct senses, including vibration, stretching, pressure, and thermoception, or the sense of temperature. Each of these senses has distinct receptors that respond to stimuli from the environment. Nociception, the response to noxious stimuli, alerts the brain when extreme temperature, chemical reaction, or some other danger threatens body tissue. Nociception can be triggered by a bodily injury or by something like the chemical capsaicin in a hot chili pepper. It could be argued that a person's understanding of time, chronoception, is a distinct sense. Circadian rhythms regulate activities like hormone levels and the desire to eat and sleep. Like the traditional five senses, these additional ways of perceiving contribute to building a model of the environment so that people can navigate through it.[23]

A second point about the senses is that they are inherently multisensory and multimodal. In other words, the sense of sight or hearing does not exist in isolation, and one sense can influence what people experience from another sense. A common example is the McGurk Effect, discovered by cognitive psychologist Harry McGurk and his assistant John MacDonald. If a person observes a video of someone saying the word “ga” while listening to dubbed audio of that person saying “ba,” they perceive the sound “da.” When that person closes their eyes and continues listening, the sound reverts to “ba.” The sight of the person saying a different word changes what sound is heard. The combination of senses helps to build a model of the environment while it attempts to resolve conflicts between different senses.[24]

While senses have evolved to do a pretty good job of helping humans survive, they are far from perfect. In fact, senses fool people all the time. As a principle, it is important for organisms to notice changes in the environment, so people are attuned to notice relative differences more than absolute measurements. For example, many people have a sense of relative pitch, or whether a subsequent note is higher or lower than the one that preceded it, but relatively few people have perfect pitch, or the ability to identify the note of a sound.[25] People also tend to mentally complete objects based on incomplete perceptions in an effort to identify and react to them. An animal seeing the tail of a cheetah above the tall grass would not survive if it waited until it saw the complete animal before reacting. However, this principle of incompleteness can be tricked, as when a snorkeler wears a shark fin above them in the water. The Kanizsa’s Triangle illusion occurs when three angles and circles with a 60 degree arc cut out are arranged in such a way as to seemingly make a new triangle appear, even though no such triangle exists.[26]

Like relative difference, the context in which the senses perceive can change what they perceive. The Ponzo illusion shows two parallel horizontal lines of the same length intersecting vertically oriented lines that converge at the top. The bottom line appears to be shorter than the top line, even though they are the same length. Likewise, the Ebbinghaus illusion shows two circles of the same size, but when each is surrounded by smaller or larger circles, they appear to be two different sizes themselves. Ambiguous drawings like the rabbit-duck illusion and figure-ground illusions like the face-vase show how people’s brains attempt to make sense of incomplete information and group sensations into workable categories by combining stimuli with prior information. Visually contextual cues sometimes and color contrasts can create illusory motion, and the color phi phenomenon creates the illusion of color change and movement when two dots of different colors in different positions are quickly displayed in succession.[27]

Finally, the senses are embodied, and people attempt to simulate the embodiment of others they observe. The brain's mirroring system responds to observing an action in a similar way as if the observer were performing the action themselves. Multiple experiments show that this mirroring system is biologically tuned: the premotor cortex responds to observations of other people performing tasks but not when people observe robots performing the same task.[28] In addition, the fusiform face area is a specialized region of the brain used to identify faces and the emotions they are expressing. With this process, people recognize someone by their facial characteristics, perceive the expression they are emoting, and feel that emotion themselves. Sound cues aid in understanding emotion; this is one reason why laugh tracks and live studio audiences are used in television sitcoms.

VR takes advantage of the senses beyond sight and sound to create an immersive experience. As people turn their heads, their sight receives new visual information, and their vestibular sense signals the radial acceleration and balance of their heads. Moving forward through a scene is observed both through optic flow and vestibular linear acceleration. Conflicts between these senses can create motion sickness, as when someone sees movement in the HMD but does not feel the acceleration with their vestibular sense. Another nauseating experience can be created when a user moves their hands in front of them and senses them with their proprioceptive sense but does not see them in the virtual world. To solve this problem, VR developers often animate virtual hands to create a visual perception of them that matches users' proprioceptive sense.[29]

One problem with stereoscopic vision in VR is the disconnect between parallax and convergence. Parallax images achieve the illusion of 3D by offsetting the left and right objects within the image. Objects that have a greater offset appear to be closer, as they mimic the change seen in closer objects in the natural world because of the horizontal spacing between the left and right eyes. People can see parallax at work by holding their finger in front of them at arm's length while alternatively closing their left and right eyes. The horizontal shift of their finger relative to the background is parallax. The relative size of objects, the occlusion of more distant objects by nearer objects, and convergence perspective also contribute to the perception of depth. However, in a stereoscopic image, a person's eyes are looking at a two-dimensional screen that is a fixed distance away. In the natural world, ocular muscles move the eyes to converge on closer objects. A person can test this by holding their finger in front of them again and focusing on it. As they slowly move their finger closer, they can feel their eyes straining to follow it. That strain is a result of convergence. The conflicting signals between perceiving parallax but not convergence in VR can lead to eye strain and motion sickness.[30]

Another issue is motion-induced vection. People can perceive motion visually by means of optic flow: as objects move past them, they see them travel further toward their peripheral vision and out of their field of view. When people see an object moving in front of a stationary background, they assume the object is moving, but when they see their entire field of vision moving, they assume they are the ones who are moving. Vection occurs when a moving object obscures the field of view to induce a feeling of motion in the observer. Think about someone looking out the window of a train at a station and seeing the train next to them begin to move. That momentary feeling of moving backward as they assume the moving train is stationary and they are moving is vection. Some VR experiences can induce vection, and the conflicting visual and vestibular signals can bring on motion sickness.[31]

While the complexities of perception can present challenges in producing VR experiences, the many idiosyncrasies between reality and how people perceive it make VR possible. As the technology improves, more types of VR experiences for longer periods of time will become possible. Moreover, developers continue to exploit the quirks of perception to heighten immersion and even trick what people are seeing as reality. Advances in AI will aid in this process. It will be important to set ethical guidelines to make transparent which images and sounds users see and hear in VR are projections of reality or manipulations of it.

THE FUTURE OF VR

With the emergence of any new technology, it is difficult, if not impossible, to predict the trajectory of that technology, what it will be used for, or how much impact it will have on the world. Thomas Edison had predicted that the phonograph would be used for recording and playing back dictation in offices, as well as books and speeches. When IBM invented the personal computer, it was convinced the PC would be a niche industry and thought it had scored a deal when a young Bill Gates agreed to be paid for his operating system in a per-unit licensing fee instead of a lump sum. In 1995, the astronomer Clifford Stoll predicted that the Internet would never replace the newspaper and dismissed the idea that one day people would read books on digital devices.[32] Accepting that predictions can be widely off, there are some general principles about new technologies that can guide thoughts about their future.

First, the current physical footprint of technology will evolve to become more functional for users. The first cellular phone from Motorola in 1983 cost $10,000 in today's dollars, had only 30 minutes of talk time per charge, and weighed almost two pounds, and the Compaq Portable computer was about the size of a suitcase and weighed about 28 pounds. While headsets today are vastly easier to wear than the Sword of Damocles, they still are bulky and can lead to head and neck fatigue. The Meta Quest 3, for example, weighs in at just over one pound, the Pimax Vision 8K is over two pounds, and the Apple Vision Pro is almost one and a half pounds, with an additional three-quarter pound battery connected to the HMD with a cable. However, developers are working to reduce HMDs down to the size of glasses weighing only a few grams. XR content is likely to be experienced using glasses instead of headsets by the end of the decade. More spectacularly, a company called Mojo had announced it was developing the Mojo Lens, billed as a smart contact lens. The company claimed it had the capability to wirelessly show videos, images, and AR content using a micro-LED display and solid-state batteries built into the lenses. Since this announcement, Mojo has scaled back to developing dynamic LED displays for VR headsets but plans on pursuing the commercial development of a smart contact lens in the future.[33] Some futurists have suggested that glasses or contact lenses will be replaced one day with brain interfaces. EEG headsets already are integrated into VR for some content; for example, the demonstration game The Adventures of NeuroBoy uses a MindWave EEG headset to move objects in a virtual environment.[34]

A second principle is that new media technologies first have a spectacle period before people discover their uses and aesthetics. People may think it is funny now when they hear a 1994 recording of morning news hosts talking about email like it is an alien form of technology.[35] Even more curiously, in the 1980s, a company sold the Cellular Phoney, a fake car phone to fool people into thinking the owner really owned a mobile phone because it was seen as a status symbol.[36] Like email, VR one day will not seem futuristic but rather become a mundane technology interwoven into the daily lives of people. Once the novelty of immersion wears off, the functional uses of VR will become clear.

Third, older media usually does not disappear when new media supplant them. Instead, they often take on new uses. Traditional mail did not end with the widespread adoption of email, but it did take on a different tenor, focusing instead on direct mail advertising and package delivery of online orders. AM radio did not die out after the better fidelity of FM became the preferred way to broadcast music. Instead, it took on a new use as the home of the emerging talk radio industry. As VR becomes a dominant medium of immersion, film, television, and other forms of immersive storytelling will continue to operate, but the functions they take on will change. An ethics of VR must also take into consideration the changes it will have on other, older technologies.

The growth of VR will bring transformative effects upon culture. VR is both a technology and an environment; that is, it is a technology that hosts an environment of human interaction. In VR, people communicate both with machines and with humans as mediated by machines. This is not a new phenomenon: all dominant media technologies move beyond functioning as tools to become environments of communication. These environments shape the way people communicate, and they take the metaphors they build in a dominant medium to the other media they use. In other words, dominant media shape human discourse in their likeness. VR has the ability to transport mental projections of self into a non-physical space. Much like the people looking down at their phones today, people in VR are physically present in one place, but mentally they are somewhere else. More so than other forms of digital communication, VR creates this mental space on a perceptive and experiential level. This will generate new expectations for older media as experiential forms of communication and entertainment, with ethical questions about identity, experience, and autonomy crossing over to these media from VR.

This is all, of course, based on the assumption that VR becomes a dominant medium of human communication. While there are no guarantees in predicting the future, some signs point to VR is traveling on this path: six million VR HMDs were sold in 2021, and sales are predicted to grow to 14 million by 2024, with an installed base of 34 million headsets.[37] During a June 2022 interview with CNBC's Jim Cramer, Meta's CEO Mark Zuckerberg said,

> By the end of the decade, we hope to basically get to around a billion people in the metaverse doing hundreds of dollars of commerce each. . . . I think there is going to be a massive economy around this.[38]

Between its growing user base, capital investment, technological advancement, and popular conceptions of immersive media, the stage is set for VR to become the primary interface for the metaverse. The earlier that ethical guidelines are constructed and debated about VR, the more prepared the public will be when it arrives.

NOTES

1. An overview of eye tracking systems in VR can be found in Adhanom, Isayas Berhe, Paul MacNeilage, and Eelke Folmer. "Eye Tracking in Virtual Reality: A Broad Review of Applications and Challenges." *Virtual Reality* 27, no. 2 (June 2023): 1481–1505. https://doi.org/10.1007/s10055-022-00738-z.

2. Google. "Google Cardboard." *Google VR*. Accessed August 27, 2023. https://arvr.google.com/cardboard/.

3. "NYT VR: How to Experience a New Form of Storytelling from The Times." *The New York Times*, November 5, 2015. https://www.nytimes.com/2015/11/08/magazine/nyt-vr-how-to-experience-a-new-form-of-storytelling-from-the-times.html.

4. A popular manufacture of CAVE systems is Visbox: Visbox, Inc. "CAVE Systems." Accessed August 28, 2023. http://www.visbox.com/products/cave/, http://www.visbox.com/products/cave/.

5. Pimax. "8K X," June 6, 2022. https://pimax.com/pimax-8k-x/; Apple, Inc. "Introducing Apple Vision Pro: Apple's First Spatial Computer." *Apple Newsroom*, June 5, 2023. https://www.apple.com/newsroom/2023/06/introducing-apple-vision-pro/.

6. Lang, Ben. "Understanding the Difference Between 'Screen Door Effect,' 'Mura,' & 'Aliasing.'" *Road to VR* (blog), February 21, 2018. https://www.roadtovr.com/whats-the-difference-between-screen-door-effect-sde-mura-aliasing-vr-headset/.

7. Xiong, Jianghao, En-Lin Hsiang, Ziqian He, Tao Zhan, and Shin-Tson Wu. "Augmented Reality and Virtual Reality Displays: Emerging Technologies and Future Perspectives." *Light: Science & Applications* 10, no. 1 (October 25, 2021): 216. https://doi.org/10.1038/s41377-021-00658-8.

8. Stauffert, Jan-Philipp, Florian Niebling, and Marc Erich Latoschik. "Latency and Cybersickness: Impact, Causes, and Measures. A Review." *Frontiers in Virtual Reality* 1 (November 26, 2020): 582204. https://doi.org/10.3389/frvir.2020.582204.

9. Redohl, Sarah. "Vuze+ 3D 360 Camera: Review, Best Practices and How to Use It." *Immersive Shooter*, August 27, 2018. https://www.immersiveshooter.com/2018/08/27/vuze-3d-360-camera-review-how-to/.

10. For a demonstration on how easy it is to port software developed with a gaming engine to VR, see: Paul, Cassaundra. "How To Play Fortnite On VR." *CellularNews* (blog), July 21, 2023. https://cellularnews.com/mobile-accessories/how-to-play-fortnite-on-vr/.

11. To learn more about the psychological effects of the uncanny valley, see Hsu, Jeremy. "Why 'Uncanny Valley' Human Look-Alikes Put Us on Edge." *Scientific American*, April 3, 2012. https://www.scientificamerican.com/article/why-uncanny-valley-human-look-alikes-put-us-on-edge/.

12. For more examples of AI-generated images, see Heaven, Will Douglass. "Welcome to the New Surreal: How AI-Generated Video Is Changing Film." *MIT Technology Review*, June 1, 2023. https://www.technologyreview.com/2023/06/01/1073858/surreal-ai-generative-video-changing-film/.

13. Meta Platforms, Inc. "Guardian System." *Oculus Developers*. Accessed September 15, 2023. https://developer.oculus.com/documentation/native/android/mobile-guardian/.

14. FusedVR Team. "Implementing VR Scene Transitions." *Game Developer*, June 13, 2017. https://www.gamedeveloper.com/design/implementing-vr-scene-transitions.

15. Matthews, Paul. "The Future of VR and AR." *IEEE Computer Society*, May 1, 2018. https://www.computer.org/publications/tech-news/trends/the-future-of-vr-and-ar/.

16. Harold Rheingold has written an excellent history of VR, including recounting his experiences trying out several early prototypes, in Rheingold, Harold. *Virtual Reality*. New York: Touchstone/Simon & Schuster, 1992.

17. Rheingold, *Virtual Reality*.

18. Lanier, Jaron. *Dawn of the New Everything: Encounters with Reality and Virtual Reality*. New York: Picador/Henry Holt and Company, 2018.

19. Lanier, Jaron. "VPL Research." *Virtual Reality Society*, September 8, 2013. https://www.vrs.org.uk/virtual-reality-profiles/vpl-research.html.

20. Kickstarter. "Oculus Rift: Step into the Game," January 30, 2016. https://www.kickstarter.com/projects/1523379957/oculus-rift-step-into-the-game.

21. Hayton, Phil. "Oculus Quest 2 Sales Jump Ahead of Xbox Series X and Series S." *PCGamesN*, June 8, 2022. https://www.pcgamesn.com/oculus/quest-2-meta-sales-xbox-series-x-s.

22. Apple, Inc. "Introducing Apple Vision Pro: Apple's First Spatial Computer."

23. Schwartz, Bennett L., and John H. Krantz. *Sensation & Perception*. 2nd ed. Los Angeles: SAGE, 2019.

24. Perera, Ayesh. "What Is The McGurk Effect?" *Simply Psychology*, September 7, 2023. https://www.simplypsychology.org/mcgurk-effect.html.

25. Witynski, Max. "Perfect Pitch, Explained." *UChicago News*. Accessed September 18, 2023. https://news.uchicago.edu/explainer/what-is-perfect-pitch.

26. Sakiyama, Tomoko, Aisato Sasaki, and Yukio-Pegio Gunji. "Origin of Kanizsa Triangle Illusion." In *Soft Computing in Machine Learning*, edited by Sang-Yong Rhee, Jooyoung Park, and Atsushi Inoue, 273: 95–103. Advances in Intelligent Systems and Computing. Cham: Springer International Publishing, 2014. https://doi.org/10.1007/978-3-319-05533-6_10.

27. For an examination of the perceptual mechanisms that enable these visual illusions, see: Parrish, Audrey E. "Visual Illusions: Insights from Comparative Cognition." In *Comparative Cognition*, edited by James R. Anderson and Hika Kuroshima, 15–30. Singapore: Springer Singapore, 2021. https://doi.org/10.1007/978-981-16-2028-7_2.

28. Kemmerer, David. "What Modulates the Mirror Neuron System during Action Observation?" *Progress in Neurobiology* 205 (October 2021): 102128. https://doi.org/10.1016/j.pneurobio.2021.102128.

29. Pyasik, Maria, Gaetano Tieri, and Lorenzo Pia. "Visual Appearance of the Virtual Hand Affects Embodiment in the Virtual Hand Illusion." *Scientific Reports* 10, no. 1 (March 25, 2020): 5412. https://doi.org/10.1038/s41598-020-62394-0.

30. Lin, Chiuhsiang Joe, and Susmitha Canny. "Effects of Virtual Target Size, Position, and Parallax on Vergence-Accommodation Conflict as Estimated by Actual Gaze." *Scientific Reports* 12, no. 1 (November 22, 2022): 20100. https://doi.org/10.1038/s41598-022-24450-9.

31. Luu, Wilson, Barbara Zangerl, Michael Kalloniatis, and Juno Kim. "Effects of Stereopsis on Vection, Presence and Cybersickness in Head-Mounted Display (HMD) Virtual Reality." *Scientific Reports* 11, no. 1 (June 11, 2021): 12373. https://doi.org/10.1038/s41598-021-89751-x.

32. Stoll, Clifford. *Silicon Snake Oil: Second Thoughts on the Information Highway*. New York: Doubleday, 1996.

33. Heater, Brian. "After Losing Sight of Its Initial Pitch, Mojo Vision Eyes Pivot with $22.4M Raise." *TechCrunch*, April 5, 2023. https://techcrunch.com/2023/04/05/after-losing-sight-of-its-initial-pitch-mojo-vision-eyes-pivot-with-22-4m-raise/.

34. "A New (Virtual) Reality: The Rise of VR Headsets & EEG." *NeuroSky*, April 28, 2016. https://neurosky.com/2016/04/a-new-virtual-reality-the-rise-of-vr-headsets-eeg/.

35. TODAY.com. "'What Is Internet?' Katie Couric, Bryant Gumbel Are Puzzled." Accessed September 3, 2023. https://www.today.com/video/-what-is-internet-katie-couric-bryant-gumbel-are-puzzled-62308421624.

36. NBC News. "1980s Flashback: When Cell Phones Were a Novelty." *NBC-News.com*, June 16, 2015. https://www.nbcnews.com/watch/flashback/1980s-flashback-when-cell-phones-were-a-novelty-465541187745.

37. Kirkcaldy, Andrew. "How Many Oculus Rifts Have Been Sold?" *PC Guide*, August 19, 2022. https://www.pcguide.com/vr/faq/how-many-oculus-rifts-have-been-sold/.

38. Novet, Jordan. "Mark Zuckerberg Envisions a Billion People in the Metaverse Spending Hundreds of Dollars Each." *CNBC*, June 22, 2022. https://www.cnbc.com/2022/06/22/mark-zuckerberg-envisions-1-billion-people-in-the-metaverse.html.

Chapter 2

Applying Ethics to VR

Ethics is more than doing the right thing. When someone makes decisions based on an ethical interest, that person commits to adhering to a set of norms that are transparent and understood by all parties affected by those decisions. Personal feelings of right and wrong may guide moral behavior, but individual biases and social conditioning do not make these feelings a reliable foundation for ethics. Morality and ethics tend to occupy different domains: the former for individuals, families, and religions; the latter for institutions, nations, and societies. Adhering to an ethical code is a compact between the members of a group. Fundamentally, ethics is about trust, a promise that those with power will act within the bounds of a set of rules to which the group has agreed. A violation of that trust makes it more difficult to build relationships with that group and establish social bonds that lead to public cooperation.

Formulating ethical decisions is often difficult. Beyond making choices that may require personal sacrifice or going against a person or group's self-interest, many issues are complex and have no easy answers. A system of ethics involves reasoning about motivations, risks, and likely outcomes. Even with a logical approach to solving ethical dilemmas, different contexts can cloud moral wisdom. What may seem to be an ethical choice for one group in a particular situation may not seem so judicious in another. Is a VR experience that relies on perpetuating stereotypes or that otherwise exploits a group ethical if the profits from that experience go to help members of that group? Should governments be empowered to mine VR data to seek out lawbreakers? Should VR environments be built to preserve anonymity or bind people to verifiable identities? The answers to each of these questions depend on details within individual cases, and different ethical systems may arrive at different conclusions in each of these cases.

Fortunately, an ethics of VR does not have to be constructed from the ground up. While moral philosophers focus on questions about how to live and treat one another, applied ethics focus on the much more practical goals of minimizing harm and producing positive outcomes in different professional, scientific, technological, and industrial contexts. Practical ethical recommendations for media and technology can be applied to the metaverse, but VR does present some unique characteristics that require careful consideration and modification of those principles. Before discussing the manner in which ethics can be applied to VR, it is important to first understand the extensive history of the study of ethics and the different ways philosophers have approached questions about choosing appropriate responses to moral dilemmas.

VIRTUE ETHICS

In both Eastern and Western cultures, the concept of a golden mean stretches back to the earliest philosophical traditions. Ancient philosophers framed ethics as states of being, and people are to pursue virtues as ideals of existence. Confucius taught that the Wuchang possesses five constant virtues: benevolence, righteousness, prosperity, wisdom, and sincerity. These virtues promote the ideas of acting with humanity and mutual respect, a pursuit of justice and fairness, working toward the well-being of all, and generosity of spirit and proper conduct. A respect for the virtues achieves a balance of maintaining social norms with the pursuit of good and internalizes the motivation for ethical behavior. Confucianism continues to hold influence in many East Asian cultures and throughout the world, though many have criticized the philosophy as patriarchal and encouraging fidelity to those who wield power.[1]

During the two centuries after Confucius, in another part of the world, the ancient Greek tradition under Socrates, Plato, and Aristotle developed another set of virtue ethics. While Aristotle's ethics differ from those of his teacher Plato, they share the goal of living virtuously, which is the secret to harmony and happiness. The word ethics itself derives from the Greek word "Ethos" (ἦθος), referring to the place of character or the defining characteristics of a person or culture. Ethos is one of three rhetorical appeals, according to Aristotle, along with pathos and logos. Ethos includes both the possession of wisdom and goodwill toward benevolence. Aristotelian virtue ethics is described as teleological because each object possesses its own purpose and final cause. For humans, according to Aristotle, that purpose is reason, and people are as good as they act according to reason. Aristotle argued for an education from an early age to build the practical wisdom necessary to practice with virtue, avoiding the extremes of deficit or excess while pursuing the virtue of a balanced character.[2]

Greek philosophy developed within a primarily oral culture, where the words a speaker says are inherently connected to the speaker and have a fundamental relationship with the reputation of that speaker. Technologies, too, are embedded with the values of their creators, but because they can be separated by time and space from their human inventors, people tend to ascribe intent to the technologies themselves rather than their creators. When we speak of an ethics of VR, it is important to remember that the ethical purpose rests with the creators of VR systems, the users of VR, and the people who engage with or are affected by these systems. A system of virtue ethics makes judgments not about individual actions but how those actions develop into patterns of practice that build the character of virtue. Having said that, an ethics of virtue may guide developers to act with reason, practical wisdom, and a benevolence toward those likely to make use of the technology.

DEONTOLOGICAL ETHICS: CATEGORICAL IMPERATIVES

If virtue ethics does not provide much guidance toward individual decisions beyond advising to act with balanced judgment based on reason and practical wisdom, a deontological theory of ethics places primacy on the duty to act based on ethical principles. The eighteenth-century German philosopher Immanuel Kant argued that people have a duty to adhere to immutable laws of goodwill, no matter what their natural inclination would have them do. If an ethical rule is established and agreed upon, it is the duty of everyone to follow that rule categorically, no matter the consequences. Given that individuals possess the faculty of reason, Kant insisted that using that reason to deduce the laws of goodwill can establish formal principles to follow. Because these principles are universal, deviating from them to serve individual desires or allegiances would only be a corrupting influence upon them.[3]

Like Plato's Allegory of the Cave, Kant argued that people cannot know anything beyond their experience in space and time. However, he differed from Plato by arguing that people can use their intuition, perceptions, prior experience, and empirical thought to deduce categorial moral laws through their representations by imagining them applied to themselves as well as others. If a law can be applied universally, does not contradict itself once it is universally applied, and respects the dignity of people as individual and rational beings, the command can be classified not merely as a hypothetical but as a categorical imperative.[4]

The primacy of duty to adhere to a categorical rule changes the concept of ethics from a philosophy of being to one of action. However, this system of ethics is limited in its ability to face conflicting duties. If, for example, it is a categorical imperative not to lie or be deceitful to others, are there absolutely no situations where being deceitful may lead to a greater good? Defenders of Kant argue that circumstances help to determine how to respond, as categorical imperatives of a higher order may take precedence over other conflicting imperatives.

Another criticism is the intercultural limitations of categorical imperatives. In a global environment where many different cultures with separate value systems interact, it is probable that different imperatives from each culture may come into conflict. Postmodern critics would point out that, in many situations, dominant ideologies masquerade as categorical imperatives. For example, the imperative within a patriarchal culture that a wife is subservient to her husband would come into conflict with an imperative from a more egalitarian culture that assumes both spouses possess an equivalent amount of power. It may be argued that ideology is embedded within Kant's ethics itself with its assumption that people are rational actors possessing free will and capable of reason separate from emotion. More recently, neuroscientists like Antonio Damasio have argued that emotion is biologically necessary and an inherent part of decision-making.[5] Cognitive biases like anchoring, loss aversion, source confusion, and anecdotal fallacies, in fact, make the entire notion of rational actors suspect. However, Kant's understanding of a respect for the individual dignity of people, as a marker of the Enlightenment, also serves as a basis for a rights-based ethics, as will be discussed further.

Applying categorical imperatives to VR would require adherence to universal guidelines about its development and use. In this possibility, one could posit a future when VR serves to educate and promote the well-being of its users, and those who use it have the best interests for themselves and others in mind. Beyond its existence as what many would label a naïve utopian view, questions emerge about the motivation of those who would develop such technology, as well as the vested interests of those engaged in its development. If a corporation has a fiduciary responsibility to its shareholders to build wealth, does it violate that responsibility by adhering to an imperative that may reduce engagement with its customers? When applying universal laws to virtual worlds, can those imperatives differ in moral character from those imposed in the real world? Is it proper, say, to virtually kill someone as a form of entertainment because it is not a physical act, or does the emotion associated with such an action necessarily corrupt the person performing the act? Similar debates that have long been argued about violence and video games can be applied to VR, but the higher level of immersion in virtual violence may change the character of that experience.

CONSEQUENTIALISM AND UTILITARIANISM

While categorical imperatives place primacy on the law to follow, consequentialism focuses on the outcome of the ethical choices people make. In the late eighteenth century, Jeremy Bentham wrote that it was the consequences of, not the motivation for, the actions of people that mattered. Following his predecessors John Gay and William Paley, Bentham argued that the good of the community could be measured by summing the total amount of happiness of all the individual members of the community. The utility of any ethical guideline could thus be measured by the greatest amount of good and the minimum amount of harm that guideline could produce.[6]

In the mid-nineteenth century, John Stuart Mill expanded on the principles Bentham had laid out to argue that the experience of happiness should be judged by the extent to which happiness is desired. While Mill argued that the quality of happiness should be factored into balancing the measurement of the quantity of happiness among people, the appeal of utilitarianism rests in its promise that its effects can be quantified to promote a net increase in happiness among a population. That quantity may be measured as a probability of a happy outcome rather than an absolute quantity of happiness and the relative differences between choices that may bring a happy or unhappy outcome. The appeal of such a philosophy in the Enlightenment, with its allusions to democracy, is in its counting a plurality of happy outcomes as desired by the greatest number of people.[7]

While practicing utilitarianism may seem to support a fair and democratic system of ethics, it ultimately falls short in protecting the rights of those who have a different value system from the majority. This is something that James Madison recognized when he warned about the "oppressive combinations of a majority" undermining the pursuit of justice.[8] A more fundamental problem with the philosophy is the limits of measuring the quantity of happiness. For example, if the early deaths of a few people, say, from air pollution, would also produce a small increase in happiness among many more people from the energy used that had produced the pollution, at what point is the utility of the many greater than the suffering of the few? Is the desire of cheap energy by a multitude of people worth sacrificing the lives or health of those harmed by the extraction and burning of fossil fuels? Conversely, is there no limit to expenditures in the effort to save a life?

Utilitarianism has a particular appeal in capitalistic digital cultures: if happiness can be quantified, it can be converted to data, which is easily modeled and processed with enough computing power. However, converting human desires into data can distort or hide inconsistencies in modeling the data, such as sampling bias or the Hawthorne effect, where people change their behavior because they are aware of being measured. Gerrymandering is another example of data manipulation: conflicting values can be grouped together in different ways to hide inconsistencies and present a misleading consensus about an ethical decision.[9]

A utilitarianism ethics of VR may provide measurable outcomes to promote the most beneficial uses of the technology, but several questions escape the scope of this philosophy, like whose happiness should be privileged over others. Should the desires of core users drive the decisions of VR's development, or should protections be put in place for casual users so that they are not harmed by the choices of developers or core users? Are there certain rights people should be granted in virtual worlds? In the future, when interacting with government or business services requires a virtual presence, can people demand freedom from using VR? This last scenario may seem implausible, but consider the number of services that currently require an online account or access to the Internet to effectively work and communicate. At the end of 2021, President Biden signed an executive order directing several government agencies to move their services, including the ability to access government benefits, online.[10] Even when offline options of accessing services are possible, the greater convenience of online access places a tax on those without Internet access. Given the cost savings of meeting in virtual environments over transportation to and investment in physical buildings, there may emerge a utilitarian pressure to require VR for many activities.

DISCOURSE ETHICS

Arriving at an ethical decision in a democratic way requires laying out and debating arguments about different positions and their possible effects. Deontological ethics focuses on the duties of ethical actors, while consequentialism places emphasis on the outcomes of ethical decisions. However, making ethical choices requires negotiating possible actions. Jürgen Habermas revises Kantian ethics by reframing ethics as a process involving mass participation and self-determination. For Habermas, ethics emerges from rational discourse and arriving at a consensus about decisions.

Using Lawrence Kohlberg's conception of developmental psychology, Habermas argues that moral development from childhood to rational adulthood can be universalized based on practical experience and the ability to hypothesize different outcomes.[11] A just ethical decision, according to this theory, is one in which the interests of everyone affected by the decision are satisfied and can freely accept it.[12] This argumentative form of ethics requires rational actors who are capable of adopting both participant and observer roles in communicating and understanding the wants and needs of others. It is this "reciprocity of behavioral expectations" that allows consensus about decisions.[13]

Like Kant, Habermas considers ethics as a matter of duty or obligation, but his theory of justice rests in the outcome of rational argument and listening to other arguments for a particular course of action instead of an intrinsic sense of right or wrong.[14] Socially normative behavior emerges from the daily practice of living and observing the behavior of others, leading to moral judgment through social cognition. While values are a priori, or deduced from reasoning without first experiencing a situation requiring their application, social norms emerge out of practical, everyday experiences.[15] Discourse ethics are not handed down from an authority but emerge from the people that are affected by them.

The limitations to this theory rest on the assumption of the existence of a civil society. People are not always rational actors, and the choices they make can sometimes be to their own disadvantage. Habermas argues that, for discourse ethics to work, people need to act toward consensus and not from a competitive position that employs deception. However, people may act irrationally in many other ways that disrupt discourse. The everyday experiences from which social norms are built are increasingly shaped by digital media. Some actors have used social media to spread propaganda or misinformation, using non-rational appeals to motivate groups of people to take some action. While Habermas would argue that this is an example of competitive behavior, not consensus, more recent examples suggest that unintentional misinformation is spread by the structure of platforms themselves. For example, researchers demonstrated that the rabbit hole of misinformation on YouTube is guided by its algorithm, which directs viewers from sometimes innocuous videos to more extreme clips containing false information, without viewers directly searching for those videos.[16] In other cases, students have unintentionally built a panopticon in schools with Instagram accounts, where they post pictures of fellow students sleeping, slouching, or daydreaming in class. Students who run these accounts have claimed they are just having good fun, but another student represented a common sentiment when she commented, ". . . it keeps everyone on edge."[17] Both examples show how digital media can influence social perceptions and behaviors that suppress reflection and rational deliberation, two conditions Habermas identifies as necessary for discourse ethics.

The enthusiasts who make up the early adopters of VR likely are favorable to practicing discourse ethics. With Oculus emerging from a Kickstarter campaign and programmers sharing VR code on sites like GitHub, the VR developer environment has been a crowdsourced effort, even though one company now dominates the hardware market. VR community members regularly participate in online communities on several forums, including Facebook groups, Discord servers, and Steam discussions, where they engage in ethical debates. Several academics who study and work in VR have developed model codes of ethics to follow.[18] Like the early development of the Internet and the current open-source software movement, people developing VR standards are working to keep platforms available to independent developers and users, even persuading Meta to announce that its work on the metaverse will not be a closed system. "Like today's internet, the metaverse will be a constellation of technologies, platforms, and products," Nick Clegg, president of global affairs for Meta, announced in 2022. "It won't be built, operated or governed by any one company or institution."[19] The VR community has formed a public sphere of communication, using Habermas' framework, to argue for structural and behavioral guidelines to benefit the members of that community. As the VR industry grows and more people participate in the metaverse, it remains to be seen whether discourse ethics will continue to take into account the effects of decisions on all users or if some people will be disadvantaged by rules negotiated by those with more power.

HUMAN RIGHTS

Discourse ethics provides a democratic system of debating rules that serve to benefit the public, but there may be some basic rules that people have agreed to be beyond debate. The concept of human rights that have been granted by a deity can be traced back to sacred religious texts, but in the seventeenth century, Thomas Hobbes and John Locke established the concept of a social contract between a ruler and the ruled and the establishment of natural rights.[20] The U.S. Declaration of Independence refers to those rights in its preamble as "certain unalienable Rights."[21] The ethics of respecting those rights are often asserted as categorical imperatives, but rights often conflict with one another when put into practice: Does the right to religious expression, for instance, allow for gender discrimination, or when does the right to own property impede the right to humane living conditions for others? These competing human rights bring up ethical questions about their implementation, hierarchy, and universality.[22]

In 1948, the newly formed United Nations General Assembly passed the Universal Declaration of Human Rights. This document enumerates rights agreed to be applied universally to all persons, including the right to personal security, privacy, property, education, and freedom of movement.[23] While governments are permitted to limit these rights to promote "public order and the general welfare in a democratic society,"[24] the declaration states that they may not do so arbitrarily or for the purpose primarily of eliminating any rights. The document, produced as a response to the Holocaust and other horrific atrocities that were committed during World War II, is Western-centric and favors democratic, representational forms of governance. Serving as a moral guide to the rule of law, the declaration is often cited to promote individual freedom from government persecution and to support the prevention of human rights abuses.

This declaration continues to serve as a guide for international law and individual rights for citizens residing within most member states, but does it also apply within the metaverse? One could argue that laws derived from the declaration apply to participants who reside in the physical location where those laws apply, but what should govern interactions between people who live in different places? Do actions performed in VR take place in that virtual environment or where a person is physically wearing their headset? If a participant engages in conduct that violates laws in the countries of other participants but not their own, should they be held to account? Do virtual worlds require their own laws separate from those that apply in physical territories? If so, should those laws be determined by private parties such as the creators of the spaces, participants in them, or some other public group? These questions arise because VR enables copresence through the simulation of physical space. Like an international telephone call or email thread, virtual worlds can connect people from around the world, but the immersive nature of VR perceptually generates the feeling that participants are in the same place.

More broadly, the actions that take place in virtual worlds complicate developing an ethics of human rights. VR is often associated with imaginative environments, like the levels of a video game inhabited by characters who may be controlled by other humans or computer algorithms. The actions that people perform while playing videogames are not considered real—a player is not *really* shooting aliens or crashing a car—but their behavior toward other people in the game can have real consequences. The hypothetical nature of putting on personas or acting out deviant behavior may be safer, or even educational, in a play setting as offered in gaming. In fact, fiction often provides lessons on social norms by showing characters who violate those norms and the consequences they face. However, as people interact with other real people within VR, the line between imaginative play and reality blurs. An application of human rights to VR may discourage antisocial behavior, unequal treatment of participants, and persecution of users, but it may also impede the free expression of those testing out the boundaries of acceptable actions.

RACE AND TECHNOLOGY

The promise of equal treatment within a technological environment governed under an ethics of human rights shows cracks when examined through the lens of race. Attempts to universally apply human rights often incorporate social assumptions of a culture that obscure misrepresentations, historical oppression, and unequal access to resources. Digital technologies replicate and amplify socially constructed racial biases and make them opaque by presenting seemingly neutral tools that make decisions more efficiently. Ruha Benjamin's analysis of race and technology has led her to declare this period as a "New Jim Code," or an environment where shopping, credit scores, job applications, and even names themselves are racially coded.[25] AI algorithms are trained on data sets imbued with these historically established values and associate an increased value to whiteness. This leads to people of color facing the choice of having to erase their Blackness to participate fairly or risk persecution by these algorithms.

In *Algorithms of Oppression*, Safiya Umoja Noble has documented how search engines produce different results depending on the race of the user and the race being searched. For example, a Google search of "Black girl" returns sexualized images much more so than "White girl," and facial recognition software identified images of African Americans as apes. This algorithmic oppression, according to Noble, reveals the problem of relying on machines to ingest and automate processes when the data they are modeled on embed inherently racist messages.[26] These technologies incorporate Black experiences of racism into the code itself, replicating and amplifying stereotypes and dehumanizing language.

The utopian rhetoric about the metaverse espouses a global village where everyone can participate in VR. A selling point of the technology, in fact, is the ability to meet anyone from around the world in a virtual environment. Putting aside the digital divide issue, in which VR is available only to those with access to a headset and a fast Internet connection, the *laissez-faire* philosophy of permitting people to be free to appear and act how they like in VR likely perpetuates racial biases in these environments. Claims of equal treatment hide the color-blind racism that denies the recognition of negative treatment of people of color in VR, which in turn prevents addressing those issues.[27] Moderation of virtual spaces also requires awareness of conscious and unconscious biases that can influence who is treated with more scrutiny. Without such care, policing in the metaverse may mimic that in physical neighborhoods of color, punishing non-white participants while favoring a culture of whiteness.

An important difference between virtual and physical worlds is the ability to change one's appearance in VR. Anyone can choose an avatar with lighter or darker skin, or choose to be an alien, or a robot. While some people may argue that this freedom allows for participants to mask their identity and avoid the pitfalls of negative treatment based on race, masking one's race does little, if anything at all, to eradicate racial discrimination. First, race is coded in language and speech. Idioms, slang, and accents often reveal, if not the race of the speaker, the mode of language being used by that speaker. Many multiplayer VR environments generate animated avatars to visually represent participants but mix in their real voices similarly to a group audio chat. Second, the avatars that users choose to represent themselves, and how are they treated based on those representations, perpetuate racial disparities. Builders of virtual environments may consciously or unconsciously embed racial codes into these environments, where users are unknowingly (or, more disturbingly, willingly) guided toward discriminatory behavior. Alternatively, VR may offer new and creative ways to explore race and transcend structural barriers to engage in critical discussion of race.[28]

ETHICS OF GENDER

When examining cultural constructs that require revision to support expanding notions of equality, the ethics of race and gender are sometimes conflated. However, it is important to consider them separately to prevent obscuring issues that are unique to each. Sexual discrimination against women and LGBT+ communities within patriarchal structures, changing understanding of the differences between sex and gender, and awareness of them as spectrums rather than binaries have evolved discussion of equal treatment and the universality of ethics. In the 1970s and 1980s, Carol Gilligan questioned the exclusively male dataset psychologists used to determine stages of moral development. In response, she described many women's experiences as an ethics of care, in which women treat ethics not as universal principles but as something that emerges from relationships and responsibilities to others.[29] While some later scholars criticized Gilligan's theory as essentialist and heteronormative, noting that the theory does not take into account the diversity of women's experiences, many others have found the ethics of care to explain a more interpersonal psychology of ethics that could be used by both men and women, as well as document the patriarchal treatment of women.[30] Gilligan's later work with David Richards discusses the impact of patriarchy on democracy, claiming that patriarchal avoidance of intimacy, rooted in the foundations of Western culture, leads to the promotion of wedge issues, war, and totalitarianism. A culture of intimacy and empathy promotes, on the other hand, an ethics of care that preserves democratic values.[31]

Feminist scholars have revised traditional deontological or utilitarianism theories by expanding the domain of their coverage to women and people of color. They argue that past applications of ethics relied on cultural norms that had historically treated these groups as children or property without the full vestment of rights that universality claims to grant.[32] A feminist ethics exposes the assumptions of patriarchy to advocate for a more universal application of ethics. Judith Butler writes that this work requires an ethical stance of nonviolence, which they assert is not passive and individualist but an active resistance to oppression that strengthens the political and social bonds of a civil society.[33] Arguing that gender is a signifier of sex, Butler writes that the performance of gender is continuously improvised. As a founding text of queer theory, Butler's book *Gender Trouble* argues that people do not possess a unified identity and embraces the idea of intersectionality. This concept recognizes that people with multiple identities, such as class, race, and gender, are oppressed by the social recognition of one identity at the expense of others.[34] More recently, Butler has argued in *Who's Afraid of Gender?* that the concept of gender has become a phantasm used by reactionary groups to suppress or attack women and LGBT+ communities in an attempt to reverse gains of universal rights. This makes an ethics of gender even more critical to counter this oppression.[35]

As an immersive medium in which users are represented by avatars, VR seems to be well suited for foregrounding the performance of gender. Participants in virtual spaces see those environments from a first-person perspective, allowing them to experience those environments as presented to them, with other characters represented by avatars that likely conform to their intended presentation instead of their physical appearance. VR is a space where users can play with representation, taking on different visual forms that are limited only by developers' designs. VR also allows for users to see the world from another person's perspective. BeAnotherLab, a global consortium of artists and researchers, used VR technology to build The Machine to Be Another, an embodiment virtual reality (EVR) system that connects two people to generate a body transfer illusion. The research group claims that the EVR system promotes empathy and helps users overcome negative stereotypes by allowing them to embody the perspective of another person. The collective has toured the world with their machine, connecting people of different genders and cultures with EVR.[36]

While VR can be used for the performance of gender, including facilitating a sort of virtual drag, the technology can also perpetuate social stereotypes instead of promoting empathy with others. Like other cultural artifacts, people can appropriate and distort representations of gender and race as masks to be used as forms of entertainment. When men playing video games choose female avatars, for instance, they often do so for strategic advantage in the game, not to empathize with another gender. These avatars often take on an idealized or over-sexualized female form, conforming more to social stereotypes of sex and gender and less to accurate representation.[37] Unclear boundaries between fantasy and reality in virtual spaces can lead to sexual harassment and abuse, reproducing the same patriarchal structures in VR that continue to pervade the physical world. An ethics of gender in VR may involve rethinking popular understandings of gender and questioning how VR can be used to challenge and surpass current ideologies instead of reinforcing them.

COGNITIVE ETHICS

While ethics is often thought of as a political or social issue, it also has become a subject of study for the fields of evolutionary psychology, social psychology, and cognitive neuroscience. Moral cognition, the study of how the brain is involved in and shapes moral and ethical decisions, examines the relationship between environmental factors and internal brain structures and functions to explain and predict the moral decisions people make. Because applying ethical rules involves reasoning, many philosophers in the past had discounted the role that emotions play in decision-making, sometimes even treating affect as a contaminant in ethical decision-making. However, neuroscientists like Vittorio Gallese, Antonio Damasio, Joseph LeDoux, and V. S. Ramachandran, among many others, are studying how the brain incorporates emotions as an inherent component of thought and memory. The embodied brain and mirroring system allow people to simulate the actions and emotions of others, and the amygdala, two small structures within the brain's cerebrum, process emotions to produce hormonal and motor responses. Emotions in this sense are action potentials, motivating individuals to act based on a perception of their environment or some internal stimulation. An empathetic response is one in which someone simulates the emotions of another person to attempt to understand how they feel. Empathy is a socially positive behavior that is likely an evolved trait to foster trust and social cooperation among a species, increasing the chance for the survival of that species.[38]

A popular thought experiment used to study moral cognition is the Trolley Problem. In this scenario, a person imagines seeing an out-of-control trolley that will kill five people on the tracks unless the person pulls a lever that will divert the trolley onto another track where only one person is standing. If the person intervenes and pulls the lever, they will save five lives but kill one. A variation of this problem imagines that, instead of a lever, the subject is standing on a bridge over the track next to a large person. If the person pushes the large person off the bridge, they will save five lives on the track but kill the person they push. The large person is often referred to as the "Fat Man," a possible allusion to the name of the atomic bomb that was dropped on Nagasaki, Japan, that ended World War II. When faced with the decision of whether to pull the lever, most people make the utilitarian decision to save more lives, but a smaller percentage of people choose to actively push a person to their death in order to save more lives. Using fMRI to measure brain activity, researchers have found that people who choose to push the man have more activity in the brain's reasoning areas, while those who did not show more activity in regions associated with empathy. Those classified as psychopaths were the most likely to push the hypothetical man onto the tracks.[39]

Moral cognition and ethical psychology use experiments like these to demonstrate that a person's brain is not a black box from which ethical reasoning magically emerges. Instead, a combination of genetic, hormonal, neurochemical, developmental, and environmental processes influence decisions people make, including what information to attend to when making decisions. Rather than providing a guide to what decisions a person should make, moral cognition seeks to explain how and why they have arrived at their ethical decisions.

With the ability to show environments from another person's perspective, VR can help promote empathy by allowing users to embody the experiences of other people. Several companies and organizations are using VR to provide first-person experiences of disadvantaged people experiencing discrimination or social stigma. One company, Helsa, markets immersive experiences under the name Empathy VR that address issues like income inequality, political polarization, and racial and sexual discrimination, while Benefit Studios distributes a collection of VR films under the title Project Empathy.[40] Knowing the role that emotion plays in attention and memory retention, developers can take advantage of priming emotions like fear, disgust, happiness, and surprise into immersive narratives to build memorable experiences, but the intense emotions that VR can generate also give developers the power to manipulate perceptions that affect moral cognition in ways that users may not be aware of. An ethics of VR must take into account this power in order to transparently develop experiences that benefit users.

Media Ethics

Ethical issues in journalism, entertainment, advertising, and public relations often are grouped together in a category called media ethics. The power of media to shape public opinion and influence policy has been a concern long before the current media environment. One of the most well-known ethical critiques of mass media was made by FCC Chairperson Newton Minow in 1961. His "vast wasteland" speech was a call to broadcast station owners to think about the social consequences of television and the need to guide it to serve the public interest. While committing to the principle of free enterprise and opposing censorship, Minow vowed to support the development of an educational television system and hold stations accountable for their content, even threatening the withdrawal of broadcasting licenses if stations did not live up to their end of the bargain.[41] Issues like violence in television programs and video games, the use of propaganda in political advertising, censorship and access to media, and representation of different groups all fall under this area of study. Media ethics examines not only what rules should be put in place, but also who has the power to set those rules.

A popular textbook on media ethics written by Clifford Christians, Mark Fackler, Kathy Brittain Richardson, and Peggy Kreshel provides a tool to examine ethical issues that media practitioners and consumers may face. These authors rely on a method of critical reflection first proposed by Ralph B. Potter, a theologian at Harvard Divinity School. The Potter Box is represented by four quadrants, each representing one dimension of ethical analysis. The first, definition, lays out the facts about the case to consider without judgment. Second, values evaluate the different ways that the facts of the case appeal to various desires or ways of living, whether they are professional goals, social principles, religious tenets, or other motivations. Next, principles bound decisions to generalizable guidelines about ethical norms, whether they be categorical imperatives or utilitarian in nature. In the last quadrant, loyalties are examined to determine which people or groups the ethical decision-maker is serving. By making loyalties transparent, decisions can be evaluated to determine whether they are self-serving or adhering to a stated principle.[42]

The Potter Box can serve as a useful tool to analyze ethical issues that may arise in VR. Beyond clearly articulating the facts of an issue, this tool encourages inquisitors to consider the values of participants, from the communitarian ethic of early adaptors to libertarian notion of unregulated action in virtual environments to consumerism, public service, democratic participation, protectionism, or other identifiable beliefs. Applying ethical principles to a case holds decisions to rational and fair application across participants and dissuades the use of ethical labels as veils to conceal preferential or arbitrary treatment. Significantly, understanding the loyalties of decision-makers may uncover the purposes for decisions that are made: Are they to serve developers, investors, platform users, consumers, non-participants, or other constituencies that benefit from, interact with, or are harmed by VR environments? Whether using it for stand-alone experiences or interactive multiuser environments, VR as a medium influences the communication that occurs within these virtual spaces. The critical analysis of the ways in which this medium is built and used can help ensure that these constituencies avoid some of its most harmful effects.

POSTMAN'S SIX QUESTIONS

Another tool that can help reflect on the ethics of VR has been proposed by Neil Postman, an educator and social critic of media and technology. Postman argues that media are technologies that both facilitate and shape communication. He describes the adoption of any new technology as a Faustian bargain: while it provides some benefits for people, it also creates new harms. "A bargain is struck," Postman writes, "in which technology giveth, and technology taketh away."[43] The concern with this, according to Postman, is that adopters of new technologies focus on all the benefits they bring without paying attention to the new problems they introduce.

In many of his books and speeches, Postman posed a set of critical questions to ask about any new technology. These questions, Postman argued, empowered individuals by foregrounding the effects of technology. Too often, Postman said, adopters of new technologies used them because they were new. The myth of inevitable progress in a technological society infers that not using a technology is standing in the way of that progress. In his last published book, Postman formulated this inquiry as a set of six questions that everyone should ask before adopting a new technology:

1. What is the problem that this technology solves?
2. Whose problem is it?
3. What new problems might the new technology create?
4. Which people and institutions might gain economic and political power once this new technology is put to use?
5. Which people and institutions might be harmed by the introduction and use of this new technology?
6. What changes in language and social organization might occur with this new technology?[44]

Like the Potter Box, these questions prompt reflection about any new technology and its possible effects on individuals, groups, and cultures. VR is a technology that solves the problem of immersion, doing a better job than large-screen movie theaters, live televised content, or first-person video games displayed on a television or computer monitor. VR places people at the center of the action instead of having them look through a window at it, and it incorporates more senses than framed media. VR also solves the problem of accurate scientific, medical, and industrial visualization better than other media. Training simulations, prototype testing, collaboration over distance, and psychological therapy are all useful applications for VR, and its value as a form of entertainment in gaming is unparalleled according to players.

However, these are problems that not everyone shares, and using VR for applications that are better and less expensively solved by older media or interpersonal communication may introduce more problems than the technology solves. For example, holding business meetings in VR unnecessarily complicates communication that may be more easily accomplished with a phone or video call. Many students and employees have had similar experiences with Zoom during the COVID-19 pandemic: solving technical glitches ("Please turn your cameras on . . . you are muted") takes away time from accomplishing the objectives of a meeting, while the technology may not necessarily contribute information that is any more useful than a simpler tool. Meetings in the metaverse fill the visual field with animated desks and avatars, adding unnecessary information and possibly detracting from the purpose of a meeting, like accidentally turning on a cat face filter during a video call.

Because the hardware and software platforms needed to build and distribute VR experiences are complex, a few companies have led their development. Two gaming engines, Unity and Unreal, are used to create most VR apps, and one company, Meta, dominates the market in VR headset sales.[45] More widespread adoption of VR likely means that these companies will benefit while making it more difficult for smaller players to compete. Because of VR's reliance on digital processing, data storage, and transmission, these large companies are aligned with the digital technology industry and share similar interests. They likely will gain power and influence if VR becomes a more pervasive means of mediated communication at the expense of other forms of communication and entertainment. How will people adjust to meeting and knowing each other through their avatars, and what incentives and conveniences will nudge people to use VR for tasks formerly undertaken with older media? Like the shopping mall in the era of Amazon, the increased use of VR will be at the expense of other institutions. Asking Postman's six questions may help people to evaluate whether VR is the most appropriate technology for a particular use, rather than defaulting to the assumption that VR is the best choice for all purposes because it is the newest technology.

THINKING ETHICALLY ABOUT VR

By presenting a historical account of different ethical philosophies, this chapter has begun to equip a toolbox for readers to approach and analyze ethical issues and think systematically about solutions to ethical problems that may arise with the widespread adoption of VR. Balancing one's own self-interest with the good of the community and other cultures is always a difficult task. Even more difficult is attempting to predict the future outcome of a decision. The futility of forecasting the future is no excuse, however, for abdicating responsibility for decisions or basing them only on the answer to the question, "What is best for me?" Even that question produces different answers depending on the context in which it is asked: self-interest in the short term may erode the trust of others and ultimately harm the decision-maker in the long term. Finding a balance, in the Aristotelian sense, between self-interest and selfless sacrifice can be challenging. The right choice looks different when considering one's duty to an already-established ethical rule, the best outcome for the most people, or who benefits and who is harmed by the decision.

While it may be impossible to know what the best ethical choice is when faced with a dilemma, reflecting on those choices rationally with knowledge of different ethical frameworks helps to provide a more honest and transparent method of making decisions. A foundation of ethical guidelines guards against biases and improper influence while making it easier to explain decisions to those affected by them, in turn making it more likely that they accept and support the decisions being made. VR as a technology and as a medium of entertainment, communication, education, work, and leisure can take many different paths as it grows and becomes more interwoven throughout our lives. The forms it takes will depend in part on the decisions that are made about its development and implementation. Ethical reflection about those decisions may help produce immersive media that benefit more people while reducing the risk of violating rights that all people share.

The following chapters present several issues to consider about VR as it becomes more widely used. As a relatively new technology, VR has faced a few ethical conflicts, but many more have yet to occur. This is the best time to intervene with critical reflection before VR becomes entrenched as a dominant medium with structures already set in place. Because of the rear-view principle regarding new media, VR at this early stage often borrows the functions and metaphors of older media, like immersive films or VR gaming, before it settles into its own poetics. Similar ethical issues with older media, consequently, can inform future debates about VR. By applying ethical principles to both real past cases and imagined future ones in VR, developers and users can be better prepared when faced with having to make these decisions in their organizations, careers, consumer choices, and as members of an active citizenry.

NOTES

1. For a good translation of Confucius' writings on ethics, see Confucius, Roger T. Ames, and Henry Rosemont. *The Analects of Confucius: A Philosophical Translation*. Classics of Ancient China. New York: Ballantine Books, 1999. A comparison of Confucius and Aristotle and their application to contemporary issues can be found in Sim, May. *Remastering Morals with Aristotle and Confucius*. New York: Cambridge University Press, 2007.

2. Aristotle and Anthony Kenny. *The Eudemian Ethics*. Oxford World's Classics. Oxford: Oxford University Press, 2011; Aristotle, J. A. K. Thomson, and

Hugh Tredennick. *The Nicomachean Ethics*. Penguin Classics. New York: Penguin Books, 2004; Heinaman, Robert, ed. *Plato and Aristotle's Ethics*. Ashgate Keeling Series in Ancient Philosophy. Burlington, VT: Ashgate, 2003.

3. Kim, Halla. *Kant and the Foundations of Morality*. Lanham, MD: Lexington Books, 2015; Wood, Allen W. *Kant's Ethical Thought*. Modern European Philosophy. New York: Cambridge University Press, 1999.

4. For an English translation of Leo Strauss' letters in which he discusses the relationship between Plato and Kant, see Shell, Susan Meld, ed. *The Strauss-Kruger Correspondence: Returning to Plato through Kant*. New York: Springer Science and Business Media, 2018.

5. Damasio, Antonio R. *Descartes' Error: Emotion, Reason, and the Human Brain*. New York: Putnam, 1994.

6. A collection of Bentham's writing can be found here: Bentham, Jeremy, Stephen G. Engelmann, and Philip Schofield. *Selected Writings*. Rethinking the Western Tradition. New Haven: Yale University Press, 2011. A historical account of Bentham's philosophical development and application of utilitarianism to law is laid out in Schofield, Philip. *Utility and Democracy: The Political Thought of Jeremy Bentham*. Oxford: Oxford University Press, 2009.

7. Mill, John Stuart. *The Basic Writings of John Stuart Mill*. 2002 Modern Library pbk. ed. New York: Modern Library, 2002. A recent discussion of how Mill's philosophy fits within democratic practice is in Cerovac, Ivan. *John Stuart Mill and Epistemic Democracy*. Lanham, MD: Lexington Books, 2022.

8. Madison, James. "Founders Online: The Federalist No. 51, [6 February 1788]." *University of Virginia Press*. Accessed July 14, 2022. http://founders.archives.gov/documents/Hamilton/01-04-02-0199.

9. For examples of how algorithms' misrepresentation of data harms people, see O'Neil, Cathy. *Weapons of Math Destruction: How Big Data Increases Inequality and Threatens Democracy*. 1st ed. New York: Crown, 2016.

10. Lapowsky, Issie. "Biden Will Sign an Executive Order to Move Government Services Online." *Protocol*, December 13, 2021. https://www.protocol.com/bulletins/biden-executiv-order-government-modernize.

11. Habermas, Jürgen. *Moral Consciousness and Communicative Action*. Studies in Contemporary German Social Thought. Cambridge, MA: MIT Press, 2001.

12. Habermas, *Moral Consciousness and Communicative Action*, 120.

13. Habermas, *Moral Consciousness and Communicative Action*, 163.

14. Fultner, Barbara, and Jürgen Habermas, eds. *Jürgen Habermas: Key Concepts*. Durham: Acumen, 2011.

15. Habermas, *Moral Consciousness and Communicative Action*, 164.

16. Tang, Lu, Kayo Fujimoto, Muhammad (Tuan) Amith, Rachel Cunningham, Rebecca A. Costantini, Felicia York, Grace Xiong, Julie A. Boom, and Cui Tao. "'Down the Rabbit Hole' of Vaccine Misinformation on YouTube: Network Exposure Study." *Journal of Medical Internet Research* 23, no. 1 (January 5, 2021): e23262. https://doi.org/10.2196/23262.

17. Rosman, Katherine. "The Latest High School Prank? It's a Snooze." *The New York Times*, December 23, 2021. https://www.nytimes.com/2021/12/23/style/school-sleep-prank.html.

18. Two examples of VR ethical codes are: Madary, Michael, and Thomas K. Metzinger. "Real Virtuality: A Code of Ethical Conduct. Recommendations for Good Scientific Practice and the Consumers of VR-Technology." *Frontiers in Robotics and AI* 3 (2016). https://www.frontiersin.org/articles/10.3389/frobt.2016.00003; Santos, Marc Ericson. "Code of Ethics for Virtual Reality." *Inborn Experience (UX in AR/VR)* (blog), October 10, 2019. https://medium.com/inborn-experience/code-of-ethics-for-virtual-reality-ebbdc5ae8ebc.

19. Clegg, Nick. "Ensuring an Open and Interoperable Metaverse." *Meta* (blog), May 18, 2022. https://about.fb.com/news/2022/05/ensuring-an-open-and-interoperable-metaverse/.

20. Hobbes argued that people possessed natural rights to fight for survival, which leads to continuous war. Only by entering into a social contract with a sovereign could peace be established: Hobbes, Thomas. *Leviathan*. Penguin Classics. Harmondsworth: Penguin Books, 2017. Locke asserted that people have three essential rights to life, liberty, and property: Locke, John. *Two Treatises of Government*. Cambridge Texts in the History of Political Thought. New York: Cambridge University Press, 1988.

21. U.S. National Archives and Records Administration. "Declaration of Independence: A Transcription." *National Archives*, November 1, 2015. https://www.archives.gov/founding-docs/declaration-transcript.

22. A good discussion of these competing rights can be found in: Monteiro, A. Reis. *Ethics of Human Rights*. New York: Springer, 2014.

23. United Nations General Assembly. "Universal Declaration of Human Rights." *United Nations*, 1948. https://www.un.org/en/about-us/universal-declaration-of-human-rights.

24. United Nations General Assembly, "Universal Declaration of Human Rights," *Art*. 29 (3).

25. Benjamin, Ruha. *Race after Technology: Abolitionist Tools for the New Jim Code*. Medford, MA: Polity, 2019.

26. Noble, Safiya Umoja. *Algorithms of Oppression: How Search Engines Reinforce Racism*. New York: New York University Press, 2018.

27. The perpetuation of systemic racism through color-blind discourse is described in: Bonilla-Silva, Eduardo. *Racism without Racists: Color-Blind Racism and the Persistence of Racial Inequality in America*. 6th ed. Lanham, MD: Rowman & Littlefield, 2022.

28. Lisa Nakamura has explored the encoding and reproduction of racial imagery, stereotypes, and discrimination in digital cultures. Two of her books that present these ideas are: Nakamura, Lisa. *Digitizing Race: Visual Cultures of the Internet*. Minneapolis: University of Minnesota Press, 2008; Nakamura, Lisa. *Cybertypes: Race, Ethnicity, and Identity on the Internet*. New York: Routledge, 2002.

29. Gilligan's often-cited book that lays out this theory based on her research is Gilligan, Carol. *In a Different Voice: Psychological Theory and Women's Development*. Cambridge, MA: Harvard University Press, 2003.

30. Vinney, Cynthia. "Gilligan's Ethics of Care." *ThoughtCo*, June 30, 2019. https://www.thoughtco.com/ethics-of-care-4691476.

31. Gilligan, Carol, and David A. J. Richards. *The Deepening Darkness: Patriarchy, Resistance, and Democracy's Future*. New York: Cambridge University Press, 2009.

32. An excellent survey of feminist perspectives on the philosophy of ethics is provided in: Norlock, Kathryn. "Feminist Ethics." In *The Stanford Encyclopedia of Philosophy*, edited by Edward N. Zalta. Metaphysics Research Lab, Stanford University, 2019. https://plato.stanford.edu/archives/sum2019/entries/feminism-ethics/.

33. Butler, Judith. *The Force of Nonviolence: An Ethico-Political Bind*. New York: Verso, 2020.

34. Butler, Judith. *Gender Trouble: Feminism and the Subversion of Identity*. New York: Routledge, 2006.

35. Butler, Judith. *Who's Afraid of Gender?* 1st ed. New York: Farrar, Straus and Giroux, 2024.

36. More information about EVR can be found here: BeAnotherLab. "The Machine to Be Another." 2022. https://beanotherlab.org/home/work/tmtba/.

37. Martey, Rosa Mikeal, Jennifer Stromer-Galley, Jaime Banks, Jingsi Wu, and Mia Consalvo. "The Strategic Female: Gender-Switching and Player Behavior in Online Games." *Information, Communication & Society* 17, no. 3 (March 16, 2014): 286–300. https://doi.org/10.1080/1369118X.2013.874493.

38. Damasio. *Descartes' Error: Emotion, Reason, and the Human Brain*; LeDoux, Joseph E. *Synaptic Self: How Our Brains Become Who We Are*. New York: Penguin Books, 2003; Ramachandran, V. S. *The Tell-Tale Brain: A Neuroscientist's Quest for What Makes Us Human*. New York: W.W. Norton, 2011. A discussion of how media uses the brain's aptness for embodied simulation and empathy is in: Gallese, Vittorio, and Michele Guerra, eds. *The Empathic Screen: Cinema and Neuroscience*. Oxford: Oxford University Press, 2020.

39. An extended discussion of this thought experiment can be found in Edmonds, David. *Would You Kill the Fat Man? The Trolley Problem and What Your Answer Tells Us about Right and Wrong*. Princeton, NJ: Princeton University Press, 2015. For an entertaining exploration of the trolley problem, listen to the Radiolab podcast on driverless vehicles: WNYC Studios. "Driverless Dilemma." *Radiolab Podcasts*, September 16, 2017. https://www.radiolab.org/episodes/driverless-dilemma.

40. Helsa provides a description of Empathy VR at their website: https://www.empathyvr.io; Benefit Studio's VR experiences can be found at: http://www.projectempathyvr.com.

41. Minow, Newton. "American Rhetoric: Newton Minow—Address to the National Association of Broadcasters (Television and the Public Interest)." Accessed August 17, 2023. https://www.americanrhetoric.com/speeches/newtonminow.htm.

42. Christians, Clifford G., Mark Fackler, Kathy Brittain Richardson, and Peggy J. Kreshel. *Media Ethics: Cases and Moral Reasoning*. 11th ed. New York: Routledge, Taylor & Francis Group, 2020, 4–10.

43. Postman, Neil. *Technopoly: The Surrender of Culture to Technology*. New York: Vintage Books, 1993, 5.

44. Postman, Neil. *Building a Bridge to the 18th Century: How the Past Can Improve Our Future*. New York: Alfred A. Knopf, 1999, 42–53.

45. Unity: https://unity.com/; Unreal: https://www.unrealengine.com/en-US; Meta's Quest 2 headset: https://store.facebook.com/quest/products/quest-2/.

Chapter 3

Control of VR

Who should control VR? The first answer that comes to mind for most people is the most democratic: nobody. This answer usually is framed by thinking about the consumer experience of VR or the freedom of people to use the technology in the ways that they want. In general, people do not like the idea of being controlled by anyone, whether it be a governmental agency, corporation, or powerful individuals. The freedom to explore, play, and discover new uses for technologies is embedded in the fabric of an open society.

However, new technologies most often require capital for research and development, and with that funding comes expectations for returns on investment. Publicly traded corporations have a fiduciary duty to build wealth for their shareholders, and government funding for research often is motivated by supporting industries to compete with other countries. In addition, organizing a functioning system inherently requires some guardrails to prevent disreputable actors from seizing control of systems and using them to attack other participants. Building a technological system that operates well requires a degree of standardization to allow for interoperability, but who should have the right to set these standards?

Questions about control of technologies are far from new. Popular culture has, for many years, presented stories about machines spinning out of control from their human creators. The publication of Mary Shelley's *Frankenstein* in 1818 introduced Dr. Frankenstein, who harnesses the recently discovered phenomenon of electricity to reanimate human flesh, only to seek out and destroy the creature after it kills the people close to him. Made during the era of silent cinema, Fritz Lang's 1927 film *Metropolis* features a robot who takes on a human likeness and convinces workers to destroy the machines that run their city. Ultimately, the workers burn the robot at the stake. In 1968's *2001: A Space Odyssey*, a computer named HAL controls an interplanetary spaceship and kills most of the astronauts aboard in its twisted effort to carry out its human-programmed mission. More recently, an AI-powered robot named M3GAN was assigned to be a companion to its inventor's niece but attacked them instead to prevent its deactivation.

At their heart, these stories highlight anxieties about harnessing the power that technologies exert, threatening to disrupt the balance of human existence. Ultimately, the fear is that these machines will begin to do the opposite of their intended purpose and harm those that they were meant to help. A similar real-life debate is materializing around the development of AI chatbots: In May 2023, a group of AI researchers, ethicists, and technology and government leaders released a statement warning of the risk of human extinction at the hands of AI. The single-sentence statement reads, "Mitigating the risk of extinction from AI should be a global priority alongside other societal-scale risks such as pandemics and nuclear war."[1] During the same month, U.S. surgeon general Dr. Vivek Murthy issued an advisory on the negative effects of social media on children and teens, warning that the technology is supplanting face-to-face social time and sleep, ultimately impacting their mental and physical health.[2]

Like social media, VR will replace more interpersonal interactions with exchanges that are mediated by the technology as it grows in popularity. Fears expressed in movies like the Stephen King adaptation *The Lawnmower Man*, a story about a mentally challenged gardener who wields uncontrollable power in the virtual world, are often exaggerated versions of real anxieties about radical power shifts engendered by new technologies. Given that control of hardware and software development, as well as the use of VR, engender power shifts, to what degree should control of this technology rest: with users, content creators, manufacturers, or government regulators? And, without jumping to extreme cases as expressed in science fiction, to what extent should we be concerned about losing control of VR platforms to the point that the technology itself begins to exert control over the people that use it?

CONTROL BY MANUFACTURERS

VR, XR, and AR technologies require a significant amount of investment to develop and perfect the illusion of immersion for its users, limiting the number of manufacturers who can build a platform and successfully bring it to market. The business principle of economies of scale applies here: the cost of research and development can be amortized over the number of units sold, incentivizing a company to sell as many headsets as possible. In other words, if it costs a company $1 billion to develop a headset but sells only 1,000 headsets, each headset would cost at least $1 million if the company were to recoup its costs. However, the same headset could sell for $100 plus parts and labor if the company sold 10 million units. The more units sold, the more a company can finance the fixed costs of developing and improving the technology. This limits who can build the hardware and software platforms that support VR.

Another business principle creeps in, though, that encourages a single company to dominate the marketplace. Network externalities refer to the principle that that the value of a network grows in greater proportion to the number of people using that network. A platform used by a niche audience may be of limited value to those users, but as the network grows to serve a mass audience, it takes on a different character. Network externalities cause networks to transform into standards, the interface by which everyone using a service interacts with everyone else. It becomes a default, the best way for users to interact, knowing that everyone else is using the same platform.

The influence of economies of scale and network externalities tends to favor the establishment of one or a few major companies that produce and maintain that network. The Microsoft operating system and its Office applications Word and Excel became a standard for personal computing not necessarily because it was the best system but because everyone else was using it.[3] Students experience this bias whenever they attempt to turn in a paper written in Apple's Pages format. The dual influences of lower cost and a widely adopted standard reward the company that grows its consumer base faster than its competitors. These influences often lead to a single company with monopoly power over a platform. People have little choice but to use a company's products if they wish to engage with others using that standard.

An example of monopoly power over services is Alphabet's Google search engine. In 2021, an EU court ruled that Alphabet used its price comparison tool within its search engine to unfairly shut out competitors.[4] Likewise, Amazon has been accused of punishing merchants who list their products for less money on other online shopping sites, pressuring sellers to provide the best price to Amazon, and, in turn, drive even more business to the site.[5] Because most users rarely scroll past the first page of search results, top placement in those results leads to more clicks and sales. A company that controls a platform can reinforce its monopoly status by favoring its own content over competitors, ranking other companies' links lower in the results, and rendering them essentially invisible. This resembles an online version of a company town that owns the only store in that town, ensuring that its workers' wages go back to the company.

An alternative model to monopoly control over a platform is a government or private organization developing an open standard that all manufacturers and consumers can utilize. The registry of Internet domains is governed by the Internet Corporation for Assigned Names and Numbers (ICANN), a U.S.-based nonprofit organization with international participation. It was founded in 1998 after complaints from researchers that a single company, Network Solutions, wielded too much control over Internet domains.[6] ICANN is charged with the "stable and secure operation of the Internet's unique identifier systems," including the Domain Name System.[7] Despite conflicts with both state actors and criminal organizations that attempt to disrupt network traffic and routing, ICANN has for the most part maintained the stable structure of the Internet, primarily because it does not answer to any one company or government.

Like the structure of the Internet, computer software and operating systems develop in tension between centralization and distributed control. Richard Stallman launched the GNU Project in 1983 to allow programmers to share and improve upon a freely available operating system.[8] The open-source software movement grew in response to companies like Microsoft, which had dominated personal computer operating systems with MS-DOS and the GUI system Windows, as well as its Office suite of applications. The Linux operating system, Opera and Firefox web browsers, LibreOffice and OpenOffice productivity suites, VLC media player, and Audacity audio editing and mixing are all examples of collaborative efforts to produce software that anyone can modify and improve.

Immersive media technologies like VR were largely experimental until the crowdsourced fundraising of the Oculus Rift. However, after Facebook (later renamed Meta) acquired Oculus in 2016, the number of headsets sold increased dramatically, to nearly 20 million headsets by 2022.[9] In contrast, Sony's PlayStation VR2 headset sold approximately 600,000 units by April 2023.[10] Because Meta already has a globally expansive user base of over three billion people, it has the power to promote and support VR hardware and a network platform that these users can connect with one another. However, such extensive control by one company can lead to building a platform that serves the interests of that company more so than the people using it. Without a significant alternative, users are faced with the choice of accepting the terms of the dominant company to access the network it has built or not participating at all in that network.

During his presentation at the 2022 South by Southwest festival in Austin, Texas, social media innovator Mark Zuckerberg talked about his vision for a virtual world in which people use VR to interact with others. Labeling this virtual ecosystem the metaverse, he declared, "I believe that the metaverse is the next chapter of the Internet. Just like we had the mobile Internet, I think this is going to be the successor to that."[11]

Normally, predictions about future technologies are more often wrong than they are correct, but Zuckerberg controlled a company worth over $550 billion at the time, even after losing over $200 billion in value the previous month.

Zuckerberg has placed so much value on this concept that he renamed his company the previous year from Facebook to Meta. At that time, he unveiled the metaverse—virtual spaces in which people, represented by their avatars, interact with others to work, learn, and play. Meta has invested heavily in this effort, purchasing the crowd-funded VR headset innovator Oculus for over $2 billion. It also purchased Within, the premiere VR software company which produces the successful fitness app Supernatural. The fact that Within's CEO, Chris Milk, who originally described the promise of VR as "an empathy machine," sold his company to a tech giant accused of anticompetitive practices, spreading misinformation, and hiding research that its social network increases anxiety and depression among teens, was not lost on critics of the sale.

In 2004, Mark Zuckerberg created Facebook along with Chris Hughes, Dustin Moskovitz, and Eduardo Saverin in his Harvard University dorm room. After acquiring the social media photo app Instagram and bringing Facebook public in 2012, Zuckerberg purchased the messaging app WhatsApp two years later for about $20 billion.[12] With over 3.6 billion people using its products each month, Meta has built the largest digital network on the planet. How is the control of VR platforms impacted when one company controls the hardware and software supporting the virtual environment?

One benefit of Meta's dominance in the market share of VR headsets is the economies of scale it can bring to the technology. Selling over 88% of the VR headsets in the first quarter of 2022, Meta has been able to undersell its competitors by reducing its costs per headset with such a large volume of production. However, the release of the Sony PS3 VR headset has helped to reduce Meta's sales to 45% in the first quarter of 2023, and Apple's release of its Vision Pro headset may further chip away at its dominance, though its $3,500 price tag likely will not compete in the same market as the Quest.[13] Between its purchase of Oculus, installed user base of headsets, low pricing, and rebranding of the company to promote the metaverse, Meta has positioned itself as the dominant corporate player in VR.

Another benefit of Meta building such a large share of VR headset production is its ability to foster the standardization of a software platform. While Quest users can use Steam and other distributors to install software (often referred to as *sideloading*), Meta encourages the use of its own software store to purchase and install software on its headsets. Like Apple's App Store for its iPhones and other devices, the Quest Store verifies that software offered on its platform does not contain malware or other harmful code that could compromise user data, damage the headset, or produce other negative effects, such as interfering with the Guardian safety boundary that prevents users from crashing into their physical environment.

The cost of developing a distribution model for VR experiences makes it desirable to limit the number of platforms users frequent, but there is a larger benefit for software developers: the network effect. Developers want their software to be available to the largest number of users, and Meta has curated a platform to feed most VR users to developers. Meta has used the network effect in its business models for Facebook and Instagram, which offer their services for free to users in order to build large audiences that they can then sell to advertisers. The network effect is self-reinforcing: as more users are added to the platform, software developers use the platform to reach the largest number of users, increasing the number of offerings on the platform, which is more attractive for users looking for new experiences.

Apple pioneered the building of an App Store when it created a platform to distribute software for its iPhone. Rather than develop all software usable by the smartphone itself, Apple established its App Store to allow other developers to sell and distribute software to be used on its devices. By allowing other developers to contribute software to its new hardware device, the App Store grew from 500 apps at its launch to over 1.3 million apps for the iPad alone a decade later.[14] The network effect encourages developers to create software for the large market of Apple products. In turn, Apple promotes the number of apps available as a reason for consumers to purchase Apple products.

Unlike a PC, however, on which software purchased from any source can be installed, Apple allows only software available on its App Store to be installed on iPhones and iPads. Apple claims this protects the integrity of its devices against malware and unverified software that may disrupt the normal operation of its devices. However, a lawsuit against Apple brought by Epic Games, the maker of the popular multiplayer game Fortnite, alleges that the Apple App Store operates as a monopoly for the purpose of controlling and profiting from the sale of software for its devices. The CEO of Epic Games, Tim Sweeney, calls the 30% fee that the App Store charges a software tax that unfairly raises the prices consumers pay for apps, and he accuses Apple of entering into opaque agreements with large publishers and software makers that create an anticompetitive environment.[15]

Because Apple does not allow software to be installed on its devices from sources other than the App Store, it maintains control over more than the design of its hardware and the software installed on it. It also knows which users have installed what software. Though Apple asserts that it does not sell users' personal information to third parties, a lawsuit asserts that the company benefits from the information it gathers about which apps users download and use on their devices, even if users select options in the software to not be tracked.[16] Known as "server-side data," information sent from a person's device to a server is considered property of the company hosting the server. This follows the principle that the physical location of data, in part, determines who owns that data. At the beginning of the PC revolution in the 1980s, user data generated on a computer stayed on that computer or was saved on disks that could be only physically transported. Now that more information is stored in the "cloud," or data centers distributed throughout the world, ownership of data is more often governed by terms that users agree to when setting up an account with these services. This issue becomes even more complex with virtual worlds, where multiple users interact and share data on servers owned by companies hosting these platforms.

Meta's immersive platform Horizon Worlds allows multiple VR users to interact in simulated physical spaces, such as a comedy club, arenas, and concert venues, via avatars that represent each user.[17] While Meta advertises the virtual spaces as social hangouts where users can meet and make friends, it also collects data on how those users interact in those spaces. University of Virginia professor Siva Vaidhyanathan has written about the evolution of Meta as a surveillance and attention technology that nudges users to spend more time using its platforms so that it can use information about those users to attract advertisers, including political groups and public relations firms seeking to modify behavior or motivate groups to a course of action.[18] By offering its services for free to users, Meta's source of revenue comes from advertisers and data brokers, not its users. Instead, users, or more precisely, information about them, are the product it sells to advertisers.

When Meta made its offer to purchase the VR content creator Within, the U.S. federal government sought to block the sale, claiming that Meta was "trying to buy its way to the top."[19] Like the Hollywood film studios in the 1930s, there were concerns that Meta would function as a monopoly through vertical integration, a concept where a company owns all the stages of production, distribution, and sales. Vertically integrated companies have the potential to lock competitors out of participating in any part of the industry. For example, a company that owns factories that assemble cars, materials, and parts to produce them and retail outlets to sell them directly to consumers could choose to ignore bids from other parts manufacturers and dealers and rely on its own divisions, preventing competition in the marketplace. In 1948, the U.S. Supreme Court ruled in *United States v. Paramount* that studios could not simultaneously own the production companies that make films, divisions that distribute them to theaters, and the theaters that exhibit them.[20] In response, studios retained ownership of production and distribution but sold off their theater chains to other companies. Unlike the fate of Hollywood studios, Meta was allowed to complete the purchase of Within after a federal judge dismissed the lawsuit, ruling that Meta on its own does not control the entire VR industry and that other players have the potential to produce and sell content on its and other VR stores.

In some cases, however, it is beneficial for the public to grant one company an effective monopoly over a service. In the nineteenth century, competing telegraph companies merged to form an effective national oligopoly, eventually becoming dominated by a single company, Western Union. Rather than having city streets clogged with multiple cables from competing telegraph companies, a single network reduced costs and increased efficiency, while senders and receivers of messages were reassured that they could contact one another.[21] Once a decision to allow a monopoly to control an industry is made, a second decision about whether the government or a private company should hold the monopoly must be made. An editorial in 1884 compares the private monopoly of Western Union with the government's monopoly of the U.S. Postal Service, implying that the company was concerned the government would start a postal telegraph owned by the government.[22] Western Union won out, and this utility model was applied to other services that emerged later, including telephone service, electricity, and cable television.

When a government allows a private monopoly to function as a utility, it usually extracts some concessions from that company. Price controls of its services, guaranteed coverage, and public governing or regulatory boards are common checks on a monopoly's power. In the case of cable television, many cities negotiated one or more channels for educational or public access so that citizens can communicate over the medium. In some cases, these agreements included the funding of production studios and staff to help produce public-access programming.[23] Because of these concessions, it often serves a company's benefit to gain power by functioning as an effective monopoly without being declared an official monopoly or utility.

How can different ethical theories guide one's thinking through the tensions of private control of VR technologies and platforms? A virtue ethics perspective may ask whether a company is tempering its desire for growth and market share with the concern of protecting the safety and data of its users. One may ask whether a company is expanding to promote the overall growth of the industry and share the benefits of the technology or is using its monopoly status to exploit the marketplace with its power and shut out potential competitors. A virtue ethics approach can emphasize a responsible approach to VR technology, focusing on promoting the development of a corporate ethos that benefits all stakeholders, including users of its services, company employees, and the larger public, including those who are not users of the technology itself.

In contrast, a deontological theory of ethics places emphasis not on the outcomes or benefits of a company, but on the motivations and moral duty of a company's actions. From this perspective, intent matters in determining the ethical behavior of a company. Deontological ethics asserts that private companies have a moral duty to respect and treat the well-being of every user of their VR technology as an end, rather than merely a means to profit. However, companies also have a fiduciary responsibility to their shareholders to deliver profit and growth. These conflicts of universally applied categorical imperatives, the dictate to deliver growth while also serving users of VR and the larger public, are difficult to resolve. Meta has attempted to adhere to ethical practice by claiming it does not exploit user data for financial gain without explicit consent, provides equal access to its platforms without discrimination provided that users can afford the hardware to access these platforms, and has taken precautions to avoid harm to users, such as warning against potential health risks of spending too much time using its headset and moderating against the risk of online harassment in the metaverse. However, any action that would use a person merely as a means to achieve a company's goals, even if it leads to a positive outcome for people, would be considered unethical under this theory of ethics.

Applying a utilitarian theory of ethics to control over VR infrastructure seems to make more practical sense at first look, but tensions between competing interests make this difficult to identify the best outcomes for different groups. For example, serving the most amount of people to benefit from using the metaverse may require collecting data to tailor more relevant experiences for users, but this may come into conflict with those who prefer to not share information about their activity on the platform or their purchase history. In addition, moderation of virtual social spaces may be employed to minimize harassment but could also be interpreted as restricting the free exchange of expression and ideas. Algorithms that nudge users for more interaction and pro-social activity in virtual environments may be employed to promote social cohesion among participants, but they may also be interpreted as manipulation by VR companies to generate engagement in an effort to increase market share and profits. If the goal under a utilitarian framework of ethics is to create policies that result in the most good for the most amount of people, determining what each group values becomes even more important, as well as charting the long-term unintended consequences of policies. Perhaps control of VR hardware and platforms by one or a few companies can reduce costs for users through economies of scale and concentrating research and development efforts, or it may stifle innovation by allowing only a few companies to control the direction of the technology's development.

An ethic of human rights seems to be a worthy goal when regulating the control of VR, but implementing even this framework can be challenging depending on conflicting interests when determining what should be held as universal rights. A universal right to life, liberty, and security are desirable goals, but what do these rights look like in a virtual environment? Should all users have access to all virtual spaces, or can a company segregate users based on interests, past behavior, or geographic region? Does a universal right to education mean headsets should be provided for free to everyone who wishes to access educational virtual spaces? If so, is that company responsible for funding such an effort, or can it benefit financially from a mandated government policy of public investment? The claim for a universal right to non-discrimination in VR may come into conflict with cultures and religions that hold a skeptical view of this technology. A company controlling the distribution and sale of VR headsets and virtual spaces can influence who has access to this technology, limiting participation in virtual communities. A company can claim that this does not impact a person's human rights because these decisions do not impact the physical conditions of people, but as more human activity moves into virtual spaces, providing universal access to and security within these spaces likely will become an issue that impacts human rights, particularly when it comes to the right to access information that may impact a person's physical well-being.

Like the regulation of the Internet and social media, when considering how to apply ethical theories to VR to determine sensible policies regarding the control of the technology, it is important to first agree upon its categorical function: Are companies that control virtual spaces utilities or publishers? That is, do these companies merely provide a network in which users can communicate, share ideas, and build alliances and political affiliations in public virtual spaces, or are the virtual environments they create a form of speech for which they should be held accountable? If VR platforms are understood as utilities, then companies should be treated as common carriers: they should not be held responsible for what occurs in these virtual spaces but should be required to allow anyone to access these spaces in a non-discriminatory manner. However, if a company is to be understood as a publisher, like video game makers or movie studios, they could be held accountable for offending speech but should be free to choose what they publish.

As the commercial Internet developed in the 1990s, Congress provided some protections for network providers to give them room to grow. Section 230 of the Communications Decency Act of 1996 gave immunity to Internet companies for the speech they transmitted across their networks. This cleared up a legal ambiguity about whether companies could be sued if their users posted defamatory messages on their networks, like a newspaper could if it published offending articles written by its reporters or editorial staff. Though the treatment of companies in this manner seems to imply a common carrier model, the Congressional Research Service notes that Section 230 does allow for "some editorial adjustments to third-party content without being considered the provider of that content."[24] The Supreme Court recently upheld immunity as it is applied to social media platforms like Facebook and Twitter, even though these companies actively monitor content and use algorithms to curate posts and videos for users. In its decision in *Twitter, Inc. v. Taamneh*, the court concluded that by merely featuring content posted by a terrorist group, platforms did not have a responsibility for the consequences of those posts as long as they did not actively "aid and abet" criminal acts. According to the Court, a platform would have to actively support content because that content would help a criminal act to succeed, as opposed to building algorithms that happen to promote violent content because that is what engages viewers.[25]

If private companies represent virtual spaces as public squares but are allowed to control them as private property, the rules they set for these spaces, absent any ethical guidelines likely will benefit owners at the expense of users. A company may implement a code of ethics because it is the right thing to do, but there are other reasons for an owner to support a set of rules that benefit users. From a public relations perspective, an ethical code can burnish a corporate image, demonstrating to its customers that it cares about their well-being. Proposing a voluntary code can head off possible government regulation that would force the company to follow rules set by others. Finding the balance between corporate self-interest that encourages growth and the development of VR while ensuring that protections for human rights are in place is a difficult but necessary task to ensure that competing interests do not overwhelm the shared goal of a thriving and functional environment.

CONTROL BY CONTENT DEVELOPERS AND USERS

An alternative to ceding control to the companies producing VR hardware and platforms is to give control over to the users actively engaged in VR. A communitarian-based governance of VR could mirror other virtual communities. Wikipedia, for example, uses crowdsourcing to contribute content; virtually anyone can add or edit a Wikipedia entry. Volunteer editors peer review entries and moderate content posted based on different categories. These discussions are transparent and included as a part of an entry that is accessible for anyone to review. The site relies on consensus for dispute resolutions, though editors can call upon other content editors or mediators if a consensus cannot be reached. Arbitration, overseen by a group of experienced editors, is a final step in resolving disputes. Wikipedia's governing body has addressed entry manipulation and other misconduct with a series of policies that prevent gaming of this system, including banning filibustering, editing with bad faith intent, and sock puppet accounts that hide an editor's identity.[26] It is the contributions of a multitude of voluntary editors that have made Wikipedia as robust and reliable as it is today.

On the surface, user control of VR seems like a democratic solution to distributing power across a platform. However, even the most distributed networks require some degree of centralized control to support the infrastructure for their distributed architecture. The Wikimedia Foundation serves as the nonprofit body that provides physical support, including the servers and software on which Wikipedia pages rely, as well as fundraising and legal structures to support its volunteers and ensure continued operation. The foundation also engages in research for new technologies that help support the free dissemination of information, improving the integrity and accuracy of information and expanding open access to new research.[27]

One may look to existing virtual communities to understand how crowdsourcing can support distributed power in the metaverse. Philip Rosedale launched Second Life in 2003 as an expansive virtual space in which avatars could interact, work, learn, and play. The company he helped to found, Linden Lab, continues to support Second Life, which serves as a model for the metaverse.[28] Though Second Life began as a virtual space displayed on a computer screen, its virtual worlds have become more complex: over 64 million users have created accounts as of 2021, though monthly active users average around 50,000.[29] Participants in Second Life pioneered the use of virtual currency, called Linden Dollars, that can be exchanged for legal currency in the physical world, and users can apply for jobs, like virtual wedding planner or DJs, that pay in Linden Dollars.[30] Second Life offers virtual homes for sale to its users, which can be protected by a virtual security system that prevents unauthorized visitors from intruding. In an announcement about the system, an administrator from Linden Lab wrote, "Privacy and fostering an open, welcoming community can't always exist simultaneously, but we believe this new security system will allow for the best of both worlds."[31] Even a virtual community that values openness and freedom recognizes the need to balance conflicting interests.

In 2015, Rosendale moved on to launch High Fidelity, a free, open-source social VR experience. His goal was to create a free service that users could expand with their own servers. In this way, High Fidelity would function like a VR version of the World Wide Web, a distributed network not controlled by any one company; instead, it would continuously evolve as users launched more servers and visited each other's virtual spaces. Users could design their own spaces and populate them with objects they designed, or they can purchase ready-made virtual items in a marketplace supported by a virtual economy.[32]

Second Life and High Fidelity can be described as sandboxes, environments in which users provide the content. Unlike traditional video games which have objectives, a narrative, and bounded spaces, a sandbox is open to whatever its participants create. Another sandbox environment that has become familiar to many children and teens is Minecraft. Created in 2009 by Markus "Notch" Persson, a Swedish programmer, Minecraft allows players to control blocky avatars and build tools to mine resources and craft objects, including fences, doors, torches, baked goods, armor, and many other useful items. Notch founded the company Mojang, which released Minecraft on multiple platforms, before Microsoft bought the company in 2014 for $2.5 billion.[33]

Microsoft has since released more bounded versions of the game, which are called Minecraft Story Mode. However, Story Mode is not nearly as popular as another feature of Minecraft added to the game prior to Microsoft's purchase: user-hosted servers. Officially called Minecraft Realms, servers can set their own rules for participation, activity, and chatting with other Minecraft players. The website MinecraftServers.org lists over 20,000 different servers that players can log on to and play games, solve puzzles, and interact with other players.[34] According to *PC Gamer*, some of the best Minecraft servers include those that have recreated settings that resemble Middle-earth from *Lord of the Rings*, Westeros from *Game of Thrones*, and Hogwarts from *Harry Potter*, islands floating in the sky that players can "own" and trade with other players, one called "PirateCraft" that allows players to sail a ship and attack other players, and a server that is designed specifically for players on the autism spectrum.[35] One teen player, Christopher Slayton, set out to recreate the known universe in Minecraft, while educational institutions like the mechanical engineering program at Carnegie Mellon University use Minecraft to illustrate 3D concepts to students.[36] Several players have made small fortunes posting their Minecraft gameplay and tutorials on YouTube, Twitch, and other platforms, gaining tens of millions of subscribers for each channel. One such player, Sean McLoughlin, who goes by the gamer handle Jacksepticeyc, made approximately $16 million in 2018 showcasing his gameplay videos.[37]

Allowing for users to control the platforms they engage with often leads to growth and changes into forms never imagined by its developers. One unintended but positive consequence for Minecraft is the establishment of The Uncensored Library. Founded by Reporters Without Borders, this Minecraft server shows a map of countries across which players can travel. In each country, there is a collection of books containing articles that are censored in that country. Effectively, this advocacy group is hiding information in an innocuous game that people can read without detection from authorities—all that is revealed is that a person is playing Minecraft.[38] While this use of Minecraft was not originally intended by Mojang or Microsoft, these companies provided the framework that made possible this and other unintended uses outside of the control of the companies themselves.

Minecraft can serve as a model for social VR, as virtual spaces, controlled by players and individuals who set up networks of servers, enabling this crowdsourced power to grow the platform. However, these examples are difficult to compare directly with VR. One reason is the state of the technology itself: while VR headsets have greatly improved in quality, they are still in a somewhat experimental state. Headsets that provide a comfortable experience are still priced well beyond the reach of most consumers. As explored in the next chapter, an economic limitation of control involves the price of access to the medium. With most HMDs sold to consumers priced between $500 and $1,500, and the Apple Vision Pro selling for $3,500, an immersive experience is inaccessible for those who cannot afford to purchase the device used to enter these virtual environments. In addition, these devices require access to a reliable high-speed Internet connection to download software and enable multiuser experiences, another cost to the medium. Google had sought to lower the cost of access to VR with Google Cardboard, a free design that uses folded cardboard and inexpensive plastic lenses to convert the screen of a smartphone into a VR display.[39] Other manufacturers make plastic enclosures for about $20 in which users can place their smartphones and utilize their screens to produce a VR experience, but these and Google Cardboard are vastly inferior to dedicated HMDs. In effect, user-controlled VR would be limited to those who could afford access.

Another ethical issue when control of VR is turned over to users is the regulation of content provided on the platforms. When users can be represented by digital avatars that resemble real people, what is to prevent them from building and manipulating AI-powered “deepfakes,” or images that resemble other real people? For example, the VR company Union Avatars promises users that they can create full-body avatars in less than a minute. Users can upload a photograph of themselves to generate the three-dimensional avatar,[40] but nothing, other than a legal agreement users click on to gain access to the site, prevents users from uploading photographs of other people.

Meta’s Horizon Worlds, a more centrally controlled VR platform, already has had difficulty moderating the conduct of its users. *The New York Times* reporter Kashmir Hill described her experience entering the social VR platform, noting that users in a virtual space can vote to kick someone out if they find their behavior threatening or disruptive. Despite this and other protections, she discovered rooms where children interrupted adult conversations, users engaged in loutish and sexually suggestive behavior, and one user asked personal questions about the real lives of other users. Even though Meta has expressed a desire to create “almost Disney levels of safety,” according to Chief Technology Officer Andrew Bosworth, Horizon Worlds demonstrates the difficulty of creating a comfortable place for users when others have the freedom to disrupt that comfort.[41]

User polling to admit and eject other users can be a mechanism to regulate the content of user-controlled VR spaces, but similar problems have occurred on more established social media platforms that called for more centralized control. Facebook, for example, began to exercise more stringent oversight over its user-generated groups to curtail the promotion of extremist views. Meta’s vice president of engineering, Tom Alison, said at the time, “We know we have a greater responsibility when we are amplifying or regulating content."[42] Groups violating Meta's community standards would first be warned by the platform and their profiles made less visible to users, with repeated violations resulting in the deletion of the group.

Long before the rise of social media, another online social platform demonstrated the possibilities and risks of decentralized control through its users. Established in 1980, Usenet was at one time a primary online space for sharing posts organized by topic. The text-based platform allowed users to offer and receive information on topics as varied as technology, culture, history, education, popular culture, conspiracy theories, and random chats.[43] While Usenet brought people with similar interests together into online communities, it was also a virtual space where users regularly violated copyright and trademark law, uploaded cracked (pirated) software, traded in illegal goods, and engaged in flame wars. In fact, many users have noted that flame wars, arguments that take on a life of their own, often were a form of entertainment in Usenet groups. The syntax of flames sometimes resembled programming code, bracketing the offending text with: "[FLAME ON] . . . / [FLAME OFF]."[44]

Moderation of these spaces can be performed by volunteers or democratically controlled by the members of each space, but this arrangement brings up another issue when ceding control of VR to its users: Who is responsible when things go wrong? Companies or organizations that control the framework and moderate the content of a virtual space can be held accountable if someone is harmed using these products and services. Beyond the member or members who perpetrate the harm, could other users collectively be held liable for their negligence in allowing this activity to occur in these spaces? How can grievances be addressed if there is no structure to appeal to? Without moderation or other restrictions in place, a democratically controlled virtual space can devolve into illegal activity and stop functioning as originally intended by its users.

Despite its promise and progress toward photorealistic representations of virtual spaces, VR continues to be a technology in search of a "killer app," the consumer use that will drive its widespread adoption. To be sure, the technology has already proven itself worthwhile for therapeutic use and training with simulations of dangerous or inaccessible settings. With the most popular headsets priced around the cost of a gaming console, VR headsets often are marketed to gamers as a more immersive form of entertainment than traditional games played on flat screens. Apple, with its Vision Pro, is seeking to carve out a space in workplace productivity: it imagines a virtual desktop will appeal to professionals who are most likely able to afford its cost. Until VR becomes more widely adopted, manufacturers of hardware and platforms will continue to exercise some degree of control to help promote its growth.

A deontological ethics of VR may assert that it is the duty of everyone associated with the technology to cede control to users to freely use immersive environments as they see fit. Just as an individual has the right to autonomy over their physical body, so too should they have control over their virtual self and experiences. Giving users control of VR hardware and platforms respects users' autonomous choices and freedom of expression and could lead to a more democratic understanding of the benefits and harms of VR. However, a utilitarian approach may recognize the futility in assuming a positive outcome of total control by users, particularly when individual desires conflict with broader interests. For example, users invested in a more immersive experience may add features that increase the cost of devices, slowing the wider adoption of the technology. A flexible ethics of VR must provide some protections for users when their interests and those of manufacturers diverge while allowing for innovation, especially given the unanticipated ways the technology may grow.

CONTROL BY THE MACHINE

Beyond concerns about the control of VR by manufacturers, content creators, and users, an ethical framework of VR should consider how the technology itself can guide or control users. Every technology has a bias based on what tasks it performs well and what it obstructs. In the 1960s, Marshall McLuhan described how different technologies extend parts of the human body while metaphorically amputating them. For example, automobiles extend people's legs by allowing them to travel farther and faster than on foot, but they simultaneously inhibit the use of those legs while sitting in the vehicle.[45] Likewise, VR extends human vision, hearing, and vestibular systems while keeping a user's eyes focused on screens close to the face. The immersive experience is generated by an artificial environment interacting with some senses while others, like most of the haptic and olfactory senses, interact with the physical space of the user. McLuhan argued that, once a human's sensory nervous system is extended outside of the body, it can be manipulated by the technology engaged in the extension, often without notice because of the technology's function as an environment in which users are experiencing different forms of content.

VR technology shapes the content experienced through it in subtle ways. For example, while users can watch 360-degree videos of expansive terrains like outer space or under the surface of the ocean, the videos themselves require recording with a camera that predetermines the vantage point of the viewer (3-DoF). Viewers can turn their heads to see any perspective from that point, but they do not have the ability to move to another spot themselves. A

360-degree video offers viewers more freedom than traditional two-dimensional framed videos that determine a point of view, but VR video producers have learned that they cannot move cameras excessively without inducing nausea in viewers. Specifically, moving and handheld shots are not desirable, and many producers simply default to a static camera located in the center of a space.

Users can move through an immersive environment in 6-DoF VR experiences, but producers are still limited by the physical space in which users engage with these experiences. The Meta Quest headset allows users to define a playable area where they can move safely; the HMD disrupts the virtual experience and displays the physical space when users step out of that area. Designers have overcome this limitation by allowing users to use a joystick controller to indicate where to teleport or float to another virtual space, but this again is more likely to induce nausea in viewers because of the mismatch between the optic flow indicating acceleration while the vestibular sense does not. The influences of limited space and shorter experiences to prevent fatigue in VR users guide the type of content that is produced for the medium: it is easier to make a short VR video or immersive game of limited duration than use VR for an entire eight-hour workday or simulation of a leisurely country bike ride. Given that the medium is still in an attractions phase, or a moment when experiencing the immersion of VR is what attracts many users to it, short sessions within a limited physical space are not much of a deterrent. However, as the technology matures, these limitations may prevent the types of stories or simulations that require more time to develop.

Another type of control that the technology itself exerts emerges from the software used to create many of these experiences. Many 6-DoF experiences are constructed using gaming engines such as Unreal Engine or Unity. This software originally was developed to build video games played on PCs, gaming consoles, or mobile devices. The popularity of first-person perspective video games requires these engines to render spaces in three dimensions that can update in real time based on a player's actions. This makes these engines well suited for VR development: a three-dimensional image can be fed to the left and right eye displays in a headset, and the head movements of the user can control the virtual camera in the software that determines the point of view within the virtual environment. However, developers use these engines primarily to build games, and many of the functions within the software are designed for that purpose. While training simulations, design testing, work collaboration, or therapeutic treatments may prove to be more useful applications of VR, most applications distributed on Steam and the Quest Store are games. Many HMDs are priced at rates comparable to game consoles, and one headset, the Sony PlayStation VR, is specifically designed to work with a gaming console, reinforcing the association between VR and video games.

This technical foundation likely encourages the marketing of VR as a gaming platform while subtly suppressing other uses.

Even when VR is used to tell stories to viewers with 360-degree videos in the style of traditional films, VR offers viewers control over perspective with the ability to look around radially from a single point in space. Viewing is occluded not by a frame but only by objects in the space. What does this mean for the types of stories that can be told in this medium, and what is difficult to communicate through it? With traditional films, directors and editors build suspense and shape the tonal quality of scenes by withholding and revealing information at specific moments. Juxtaposition of shots, variation of shot pacing, and camera movement all help construct the meaning of the narrative. Without a rectangular frame, VR emphasizes immersion over concealment, so immersive stories tend to rely more on directing viewers' attention through sound and action over the overt control of gaze with shot distribution. Often, a VR storyteller uses sound as a directional cue to have viewers turn their heads toward where the storyteller wants them to look. VR stories place viewers within a scene as opposed to being an outside observer to the action. Some viewers have reported feeling complicit in the action they are witnessing, and VR storytellers have taken advantage of this feeling. In *Hunger in L.A.* (2012), Nonny de la Peña places viewers in the middle of an altercation at a food bank in Los Angeles. Fights break out around the viewer, who is helpless to act or calm the violence.[46] The immersive technology allows storytellers to generate this feeling among viewers, encouraging them to tell stories that implicate viewers within a story as passive participants rather than impartial observers.

An ethics of VR rightly should address the monopolistic control of creators and the safeguards against unethical control by users abusing the platforms, but the forces exerted by the often-overlooked underlying technological influences of VR should also be considered when developing policies. While manufacturers produce VR technology, content creators provide experiences and environments for users, and users engage with the platform, the technology itself can exert control by placing limits on standards, manipulating behavior while users use VR, and influencing cultural and ethical norms through its underlying structures. Like other media technologies such as television, the smartphone, and the Internet, VR has a transformative power that extends beyond its primary functions of delivering content and connecting users.

Whether VR can function, as early adopter Chris Milk has suggested, as an empathy machine, it does have the power to metaphorically transport users to the center of different cultural landscapes and simulate the point of view of another person. An ethics of care most closely aligns with Milk's empathy goal, striving for an understanding and compassion for others. The paradox of this goal is that, while VR simulates other perspectives, users are closed off from the immediately accessible physical world, their eyes masked by an HMD. Like the problem of social media activism, critics have questioned the level of change that can be achieved when viewing the circumstances of others from the comfort of their safer physical spaces. A duty ethics may better address the obligations of leveraging the technological biases of VR, focusing on its ability to represent other people and cultures truthfully without distorting those representations for entertainment value or to reinforce stereotypes. Similarly, a utilitarian ethics can strive to find the most good for which the medium can be put to use, given these tendencies of the technology to both transport users to and close them off from experiencing other people and cultures.

A discussion about control is ultimately about power. The fair distribution of control between developers and manufacturers, users of VR, and the technological biases of the medium itself is critical for equitable experiences that navigate the spectrum between security and freedom, growth and fair access, and awareness of how technical influences shape use. Policies have to find a balance that does not prevent an industry still seeking widespread acceptance from imploding without giving a few large manufacturers the power to lock out competitors or ignore the well-being of users. Finally, an ethics of control over VR must consider its impact on the well-being of people who do not use the technology. A balanced approach must consider broader societal implications, recognizing that the ripple effects of VR technology extend beyond the immediate stakeholders. An ethics of VR must navigate between fostering innovation and ensuring accessibility, safeguarding privacy without stifling creativity, and promoting responsible usage while acknowledging diverse cultural values. In doing so, an ethics of control must integrate multiple perspectives, including those often marginalized, to create a robust and inclusive framework that supports the responsible evolution of VR technology. Collaboration between groups with different interests can develop a code of ethics that is aligned with shared values and interests.

NOTES

1. CAIS. "Statement on AI Risk," May 30, 2023. https://www.safe.ai/statement-on-ai-risk.

2. Office of the Assistant Secretary for Health. "Surgeon General Issues New Advisory About Effects Social Media Use Has on Youth Mental Health." *HHS.gov*, May 23, 2023. https://www.hhs.gov/about/news/2023/05/23/surgeon-general-issues-new-advisory-about-effects-social-media-use-has-youth-mental-health.html.

3. A good history of how Microsoft founder Bill Gates dominated the competition to develop the operating system for personal computers can be found in Manes, Stephen, and Paul Andrews. *Gates: How Microsoft's Mogul Reinvented an Industry*

and Made Himself the Richest Man in America. New York: Simon and Schuster, 1994.

4. Aulner, Francois, and Foo Yun Chee. "Google Loses Challenge against EU Antitrust Ruling, $2.8-Bln Fine." *Reuters*, November 10, 2021. https://www.reuters.com/technology/eu-court-upholds-eu-antitrust-ruling-against-google-2021-11-10/.

5. Ovide, Shira. "The Big Deal in Amazon's Antitrust Case." *The New York Times*, May 25, 2021. https://www.nytimes.com/2021/05/25/technology/amazon-antitrust-lawsuit.html.

6. ICANN. "ICANN History Project—ICANN." Accessed April 16, 2024. https://www.icann.org/history.

7. ICANN. "Bylaws for Internet Corporation for Assigned Names and Numbers: A California Nonprofit Public-Benefit Corporation—ICANN." Accessed April 16, 2024. https://www.icann.org/resources/pages/governance/bylaws-en/#article1.

8. Stallman, Richard, and Richard M. Stallman. *Free Software, Free Society: Selected Essays*. Edited by Joshua Gay. 1st ed. Boston, MA: Free Software Foundation, 2002.

9. Heath, Alex. "This Is Meta's AR / VR Hardware Roadmap for the Next Four Years." *The Verge*, March 1, 2023. https://www.theverge.com/2023/2/28/23619730/meta-vr-oculus-ar-glasses-smartwatch-plans.

10. Baker, Harry. "PSVR 2 Sold 600k Units Up to April, Outpacing Original PSVR." *UploadVR*, May 24, 2023. https://www.uploadvr.com/psvr-2-600k-units-sold/.

11. Rubio-Licht, Nat. "The Metaverse Is on Its Way, According to Mark Zuckerberg." *Protocol*, March 16, 2022. https://www.protocol.com/mark-zuckerberg-metaverse-sxsw.

12. For a comprehensive history of the company, read Levy, Steven. *Facebook: The Inside Story*. New York: Blue Rider Press, 2020.

13. IDC. "IDC—AR & VR Headsets Market Share." Accessed April 17, 2024. https://www.idc.com/promo/arvr.

14. Apple, Inc. "The App Store Turns 10." July 5, 2018. https://www.apple.com/newsroom/2018/07/app-store-turns-10/.

15. Owen, Malcolm. "Epic vs. Apple: Fortnite, History, Court Ruling, Appeal Results." *AppleInsider*, April 24, 2023. https://appleinsider.com/articles/20/08/23/apple-versus-epic-games-fortnite-app-store-saga----the-story-so-far.

16. Wuerthele, Mike. "Apple Getting Sued over App Store User Data Collection." *AppleInsider*, November 12, 2022. https://appleinsider.com/articles/22/11/12/apple-getting-sued-over-app-store-user-data-collection.

17. Meta Platforms, Inc. "Horizon Worlds: Virtual Reality Worlds and Communities." https://www.meta.com/horizon-worlds/.

18. Vaidhyanathan, Siva. *Antisocial Media: How Facebook Disconnects US and Undermines Democracy*. New York: Oxford University Press, 2018.

19. Feiner, Lauren. "Meta Acquisition of Within Reportedly Approved by Court in Loss for FTC." *CNBC*, February 1, 2023. https://www.cnbc.com/2023/02/01/ftc-loses-attempt-to-block-meta-acquisition-of-within.html.

20. *United States v. Paramount Pictures, Inc.*, 334 U.S. 131 (1948).

21. An engaging history of the use of the telegraph as a new technology can be found in Standage, Tom. *The Victorian Internet: The Remarkable Story of the Telegraph and the Nineteenth Century's on-Line Pioneers*. New York: Bloomsbury, 2014.

22. "A Telegraph Monopoly." *The New York Times*, May 27, 1884, 4.

23. Janes, Barry T. "History and Structure of Public Access Television." *Journal of Film and Video* 39, no. 3 (1987): 14–23.

24. Congressional Research Service. "Section 230: An Overview," April 7, 2021. https://crsreports.congress.gov/product/pdf/R/R46751.

25. *Twitter, Inc. v. Taamneh*, 598 U.S. 471 (2023).

26. "Wikipedia: Editorial Oversight and Control." In *Wikipedia*, August 12, 2022. https://en.wikipedia.org/w/index.php?title=Wikipedia:Editorial_oversight_and_control. An early history of Wikipedia written by one of its founders can be found in Lih, Andrew. *The Wikipedia Revolution: How a Bunch of Nobodies Created the World's Greatest Encyclopedia*. New York: Hyperion, 2009.

27. Wikimedia Foundation. "Our Work." Accessed July 29, 2023. https://wikimediafoundation.org/our-work/.

28. Linden Lab. "About." Accessed July 29, 2023. https://lindenlab.com/about. A good history of virtual communities and the culture of Second Life users can be found in Pearce, Celia. *Communities of Play: Emergent Cultures in Multiplayer Games and Virtual Worlds*. Cambridge, Ma: MIT Press, 2009.

29. Second Life Community. "Number of Second Life Users?" May 24, 2021. https://community.secondlife.com/forums/topic/472381-number-of-second-life-users/.

30. Second Life Community. "Inworld Employment," July 29, 2022. https://community.secondlife.com/forums/forum/312-inworld-employment/.

31. Second Life Community. "Linden Homes Security System," April 26, 2019. https://community.secondlife.com/forums/topic/436103-%E2%9A%A1-linden-homes-security-system-%E2%9A%A1/.

32. Kushner, David. "Beyond Second Life: Philip Rosedale's Gutsy Plan for a New Virtual-Reality Empire." *IEEE Spectrum*, January 30, 2017. https://spectrum.ieee.org/beyond-second-life-philip-rosedales-gutsy-plan-for-a-new-virtualreality-empire.

33. Cox, Alex. "The History of Minecraft: The Best Selling PC Game Ever." *TechRadar*, September 4, 2020. https://www.techradar.com/news/the-history-of-minecraft.

34. "Minecraft Server List." *MinecraftServers.org*. Accessed July 30, 2023. https://minecraftservers.org/.

35. Palmer, Philip. "The Best Minecraft Servers." *PC Gamer*, November 1, 2022. https://www.pcgamer.com/the-best-minecraft-servers/.

36. Rubin, April. "A Minecraft Player Set Out to Build the Known Universe, Block by Block." *The New York Times*, October 9, 2022. https://www.nytimes.com/2022/10/09/technology/minecraft-universe-developer.html.

37. Robehmed, Natalie. "Highest-Paid YouTube Stars 2018: Markiplier, Jake Paul, PewDiePie and More." *Forbes*, December 3, 2018. https://www.forbes.com/

sites/natalierobehmed/2018/12/03/highest-paid-youtube-stars-2018-markiplier-jake-paul-pewdiepie-and-more/.

38. Reporters Without Borders. "The Uncensored Library." Accessed August 1, 2023. https://www.uncensoredlibrary.com/.

39. Google. "Set up Google Cardboard." Accessed August 1, 2023. https://support.google.com/cardboard/answer/6295046.

40. Union Avatars. "Realistic Full Body Avatar Maker." Accessed August 6, 2023. https://unionavatars.com/.

41. Hill, Kashmir. "This Is Life in the Metaverse." *The New York Times*, October 7, 2022. https://www.nytimes.com/2022/10/07/technology/metaverse-facebook-horizon-worlds.html.

42. Zadrozny, Brandy. "Facebook to Crack down on Groups That Break Its Rules." *NBC News*, March 17, 2021. https://www.nbcnews.com/tech/tech-news/facebook-crack-groups-break-rules-rcna435.

43. An interesting history of Usenet written by two of its early participants. These authors argued for a free and unrestricted Usenet, though the survey responses they presented showed a desire for more control. Hauben, Michael, and Ronda Hauben. *Netizens: On the History and Impact of Usenet and the Internet*. Los Alamitos, California: IEEE Computer Society Press, 1997.

44. Buck, Stephanie. "Remembering the 90s Flame Wars: A Simpler Time of Cyberbullying." *Medium*, August 23, 2016. https://timeline.com/flame-wars-early-cyberbullying-1c509aa5ffad.

45. McLuhan, Marshall. *Understanding Media: The Extensions of Man*. 1st ed. Cambridge, Ma: MIT Press, 1994.

46. Goodman, Lizzy. "'Hunger In L.A.' Immerses Viewers In An Interactive Journalism Experience (And A Food Line)." *Fast Company*, January 31, 2012. https://www.fastcompany.com/1679530/hunger-in-la-immerses-viewers-in-an-interactive-journalism-experience-and-a-food-line.

Chapter 4

Access to VR

Inspired by working as a technician at the University of Southern California's Institute for Creative Technologies in 2011, Palmer Luckey imagined a low-cost VR headset that could be available to virtually everyone.[1] The next year, he launched a Kickstarter campaign and raised over $2.4 million to develop what would eventually become the Oculus Rift.[2] Luckey has described VR as "the final medium," which would allow anyone from any economic or social background to participate in experiences previously unavailable to them.[3] Critics have questioned Luckey's intent to pursue this goal, given that he sold his company, Oculus, to Facebook in 2014 for $2 billion.[4] Facebook subsequently rebranded the headset as the Meta Quest to match the new name of Mark Zuckerberg's company, which reflected his focus on the metaverse. Since purchasing Oculus, Meta has marketed two versions of its VR headset: the Quest 3 for $499 and a higher-quality Quest Pro for $999. Sony sells its PlayStation VR2 headset for $549, but it requires a connection to a $500 PlayStation 5 gaming console. Apple's Vision Pro headset, which the company advertises as a "spatial computer," is priced at $3,500, even more out of reach economically for many people.[5]

Beyond the hardware itself, VR requires a high-speed Internet connection to access software and engage in any social VR experience. While some local areas provide free or low-cost access to the Internet, many people cannot afford the cost of connecting. While the price of Internet service has become less expensive than in the past, about 2.7 billion people still do not have access to a connection.[6] For those who do have a connection, it is usually provided by a mobile phone service as opposed to a fixed connection.

On a more fundamental financial level, purchasing software requires some sort of access to the digital financial economy. The Meta Quest Store, for example, requires a credit card or PayPal account to make software purchases.[7] While using PayPal may seem to offer more opportunities for access, the PayPal user agreement itself requires a bank account or credit card to open an account with the service.[8] According to the World Bank, about 1.7 billion people lack access to any type of bank account, which provides access to digital financing.[9] Mobile phones have helped to reduce this exclusion, processing over $1 billion per day across 90 countries, but half of those who remain without access to the world's digital financial system include women, the poor, and those who are not participating in the global workforce. In the United States alone, approximately 5.9 million people are without a bank account. According to the FDIC, "unbanked rates were higher among lower-income households, less-educated households, Black households, Hispanic households, working-age households with a disability, and single-mother households."[10]

While VR promises immersive sensory experiences that can educate and entertain, not everyone is guaranteed equal access to this technology. If VR is to become, in Luckey's words, the final medium, how does the industry guarantee that everyone has access to this medium, regardless of their location, income, education, disability, or other barriers to VR that people face today? Are there barriers that are insurmountable, or is the ability to overcome these obstacles merely a matter of providing incentives to achieve universal access? If universal access to VR is not achievable, what experiences will people without access be left out of, especially as the network effect pushes essential services to immersive environments?

THE DIGITAL DIVIDE

Issues of access to VR as a digital technology resemble in many ways, a longer-debated concern known as the digital divide. In 1996, only three years after the release of the Mosaic World Wide Web browser, the U.S. Congress passed the Telecommunications Act of 1996. Among making many other changes to U.S. policy regarding digital technologies, Section 706 of this law specifically addressed concerns about a digital divide between people who had access to a computer and Internet connection and those who did not by encouraging the deployment of advanced telecommunications services to all Americans. The law also gave the FCC authority to establish policies that would promote competition and reduce regulation in order to encourage the growth of the digital telecommunications industry. Section 706 states:

> The Commission shall determine whether advanced telecommunications capability is being deployed to all Americans in a reasonable and timely fashion. If the Commission's determination is negative, it shall take immediate action to

> accelerate deployment of such capability by removing barriers to infrastructure investment and by promoting competition in the telecommunications market.[11]

The law specifically addressed schools, which Congress saw as a place where Internet access could reduce inequality in information. It established the National Education Technology Funding Corporation to direct funds to build Internet infrastructure. Though the law sought to distribute funding to all schools by discounting Internet services provided to them, Mark Cooper, Director of Research for the Consumer Federation of America, remarked later that year to *The New York Times*, "Without some sort of means test, the wealthy communities would benefit the most from the discounts because they could afford to purchase the most services."[12]

Three years after the passage of the law, a report by Mark Cooper and Gene Kimmelman for *Consumer Reports* concluded that the Telecommunications Act of 1996 fell short in addressing the digital divide. The report suggests that while the act was intended to promote competition and reduce regulation in order to encourage growth in the telecommunications industry, it did not do enough to address issues related to access and affordability for low-income households. The report also noted that the law had failed to promote universal service, which is essential for ensuring that everyone has access to the Internet. Rather than providing universal access to the Internet, the report concluded that the law helped to increase the digital divide.[13]

In 2001, Harvard professor of political science Pippa Norris defined the digital divide across many dimensions, including the divide between industrialized and developing nations, as well as social and democratic divides between those who seek out information and those who either cannot or choose not to, even when free access is provided. Noting that the digital divide can be charted across differences in education level, ethnicity, age, and economic status, Norris highlighted the dangers of the divide as more governments transitioned public information and debate to digital formats, excluding citizens who do not have access to this means of information sharing.[14]

These differences in access are exacerbated by social media, where algorithms reinforce habits of information seeking and engagement. For example, the Pew Research Center asked Facebook users to view their own ad preference pages on the platform. The more that people used the platform, the more they agreed with the ad preferences listed for them and the content that was displayed on their pages. These preferences were not affected by the privacy settings users had selected on their accounts, suggesting that their engagement on the platform largely determined how Facebook set users' ad preferences and selected content to display on news feeds.[15] Rather than opening new possibilities of thought or exposing users to counterclaims, alternative cultural contexts, and information useful for making rational decisions, these algorithms often reinforce stereotypes and solidify confirmation biases in order to promote engagement with the platform.

In a literature review of research on the digital divide, professors Polyxeni Vassilakopoulou and Eli Hustad document a pattern that exists between digital inequality and physical inequality related to socioeconomic resources. While access to digital technologies has become more equitable, Vassilakopoulou and Hustad caution that greater concern should be focused on the use of digital technologies that ensure "fairness and inclusiveness." For example, the design of interfaces should take into account cultural differences that may discourage use, even though those groups may have access to these tools.[16] As more information moves to become accessible online, the rhetoric that promotes the digital conversion of public resources often assumes that all people have equal access to information. However, the digital divide often mirrors access to other socioeconomic resources, including stable employment, housing, and discretionary income that allow people the time and resources to access online information. Solving the digital divide likely requires addressing more foundational socioeconomic issues like providing access to quality education, reducing income inequality, and ensuring a base level of sustainable basic living conditions for all people.

What makes matters worse is that people without access to digital resources are also themselves often the targets of digital algorithms. In *Weapons of Math Destruction*, Cathy O'Neil lays out several examples of algorithms that reinforce inequities they ingest through their training data and that become embedded in their structure. From standardized testing to employment search firms, credit rating agencies, and predictive policing, these algorithms often function at the expense of the poor and other oppressed groups and reinforce increasingly solidified economic and social strata. The people and organizations that use these big data and AI models are, in fact, the ones who make decisions, but by outsourcing the justification for decisions to a predictive model, they distance themselves from the responsibility of the decisions by claiming to be neutrally following the rational conclusion of data processing. The cold calculations of algorithms are making the decisions to discriminate according to those in power, not the humans following the recommendations or directives of the programs. Claiming that they have removed their own human biases from the decision-making process reinforces the inequities that they perpetuate and makes it even more difficult to address these inequalities.[17]

VR technology provides a level of immersion beyond what is available with desktop and mobile computing. Currently, the technology exists as a medium of spectacle, where the thrill of experiencing the immersion that connects the panoramic visual, aural, and vestibular senses is enough of a reason to put on a VR headset. As VR moves past this stage to serve other uses, its advantage as an immersive medium will supplant older technologies which attempt to service this function but do not perform as well as VR. Because VR is built upon digital technologies that have already become a part of the environment, including digital imaging, Internet connectivity, 3D software modeling and digital rendering of the real world, real-time animation, large data models, and AI, the same biases embedded in those older technologies likely carry over into this new one. Like with Internet connectivity, the digital divide in VR should bring about ethical guidance on how to overcome economic, social, and political barriers to allow anyone access to the virtual worlds made possible by this technology.

ACCESS TO TECHNOLOGY

Understanding how the digital divide has been defined as primarily an issue of access to computing hardware and an Internet connection masks other obstacles in the digital divide, but access to VR technology itself is a precursor to gaining access to virtual worlds. Even without taking into consideration these other issues, the cost of a VR headset is a primary barrier to accessing virtual worlds, making economic inequality a primary barrier to the technology. The cost of components that are needed to build an effective HMD, including high-resolution displays, fast graphics processors, high-speed wireless data transfer, and lightweight high-capacity batteries, is comparable to a premium smartphone. Because the technology is driven by private investment to develop commercial products that will eventually produce a profit, HMDs often are marketed to early adopters of digital technologies, including gamers, professionals, and large businesses conducting their own research into immersive applications. In its marketing to consumers, Meta, Sony, and HTC have advertised VR mostly as a gaming device, and the most available software apps are VR games. A cursory look at the Meta App Store reveals that the top downloads for the Meta Quest headset include the games Gorilla Tag, Beat Saber, Blade & Sorcery: Nomad, Population: One, Job Simulator, Epic Roller Coasters, and Among Us: VR. Of the 20 most popular apps listed on the site, eighteen are games, one is a platform to build VR games, and one is a VR chat room.[18] Selling VR primarily as a gaming experience allows manufacturers to harness revenue and build a committed base of users while perfecting the technology, but it also limits who is exposed to this new technology, affecting the possibilities of imagining uses that can benefit more people.

Attempting to predict possible futures of VR and the ethical issues of access to the technology may be difficult, but authors of science fiction literature already have imagined different ways in which VR will be integrated into everyday life, as well as the challenges of providing equitable experiences with immersive media. Ernest Cline's novel *Ready Player One* demonstrates the different levels of experiences users can have. In the beginning of this story, the protagonist, teenager Wade Watts, accesses the immersive environment known as OASIS via a low-quality, school-issued headset and haptic gloves. This is much like how schoolchildren today use bare-bones Chromebooks provided by schools in order to access instructional content and instead play games or use unauthorized apps. Later in the story, after winning prize money in the competition that makes up the heart of the story, Watts purchases a fully immersive haptic suit, allowing him to experience the OASIS in more vibrant detail and providing him with more skill to navigate the environment.[19] In *Snow Crash*, the 1992 novel by Neal Stephenson that coined the phrase "metaverse," users access the simulation using high-fidelity displays that project images and sounds into goggles and headphones to produce a fully immersive experience. However, the metaverse could also be accessed using lower-quality public terminals that represent users as black-and-white, blocky avatars in the virtual world.[20] A common message in both of these stories is that economic power provides for a more immersive and effective experience, much like a faster Internet connection and more powerful computers allow for an edge in gaming or financial trading. While coding skill in building avatars carries some weight in Stephenson's metaverse, the social stigma of using public terminals to access the simulation, much like the inconvenience of having to use computers with a time limit at a public library to access the Internet, demonstrates how status and privilege are constructed from economic power and property.

Even if a person overcomes the economic limitation of meeting the cost of entry to VR and has the resources to purchase a headset, related software, a subscription to an online platform, and Internet access, other barriers to entry remain for universal access to VR technology. For example, researchers Daniel Pimentel, Maxwell Foxman, Donna Z. Davis, and David M. Markowitz outline the usefulness of VR for treating mental illness, including depression, anxiety, and PTSD, but lament that its widespread commercial use is limited by the perception of VR as primarily an entertainment device, dissuading hospitals, clinics, and health insurance companies from investing in the technology.[21] VR as a medium of entertainment infers that it is a luxury good and not worthy of public investment, even though it could be employed to help solve or mitigate several issues that affect the public.

In addition to the economic and social perception issues about VR, producing enough headsets to enable the mass adoption of VR has become a challenge. Post-pandemic supply chain issues continue to affect the industry, making it difficult to source the motion sensors, graphics processors, and wireless communications chips needed to generate the illusion of immersion.

For example, in its ramp-up to release its Vision Pro headset, Apple explored abandoning its plans to use Sony as the exclusive supplier of OLED screens and branch out to multiple manufacturers in China after it was revealed that Sony could produce screens for only about one-half million headsets in the first year of production, far below the indicated demand for the product.[22] Compounding this issue is the fact that the components needed for high-performance VR requires a low failure rate tolerance. The manufacture of semiconductor chips and LED screens inherently produces a small number of faulty circuits that are usually overcome by error correction or shipping chips with a percentage of disabled circuits. However, the real-time rendering of three-dimensional images needed by VR requires these chips to perform at an optimum rate, resulting in fewer manufactured chips passing production tests.[23]

Building enough headsets to provide access to VR is difficult enough, but compounding this problem is the demand that other industries are putting on the same components. In November 2022, OpenAI, a nonprofit research company founded in 2015, released to the public ChatGPT, a pretrained generative AI large language model. ChatGPT can respond to user-supplied prompts to generate paragraphs of text, lists, poems, and computer code.[24] Less than one year later, over 79% of corporate leaders responded in a McKinsey survey that they have had some exposure to AI, and 40% plan to increase their investment in AI to take advantage of the efficiencies that generative AI can bring to their companies.[25] The explosion of financial investment in AI has created a shortage of chips to train and run the models, which each require an immense amount of computing power. AI coding in particular runs well on GPUs (graphics processing units), a microchip originally designed to offload calculations from a computer's central processing unit to produce faster and more detailed renderings of graphics.[26] Unlike the serial processing of CPUs, GPUs are designed to take advantage of parallel processing of data, making them ideal for AI training and processing of other large, probabilistic data sets.[27] Nvidia, one of the largest and most advanced manufacturers of GPUs, has begun to market its GPUs as an essential building block for AI platforms. Sales of its chips have skyrocketed and have caused a shortage of its chips that could have been used for other purposes. According to *Wired*, a single AI platform may use "sometimes 10,000 chips at a time," reducing the availability of these chips to be put to use for their originally intended purpose of rendering graphics.[28] This increased demand for microchips has diverted production from VR to AI components, further intensifying the chip shortage.[29] This chip shortage has made it even more challenging to mass-produce VR headsets.

The concerns about the power and abuse of AI have led the U.S. government to restrict the export of GPUs and other technical components to some countries. According to CNBC's Kif Leswing, in 2022, the same year as the release of ChatGPT, the United States restricted Nvidia from selling some of its chips to China "because of fears that Chinese companies could use them for military purposes or steal trade secrets."[30] Early in 2023, the United States began to consider which other countries should be prevented from accessing high-end GPUs that could be used to build AI platforms.[31] Later that year, Nvidia revealed that the U.S. Commerce Department blocked the dissemination of AI technology by placing export controls over Nvidia and AMD computer chips being shipped to several countries in the Middle East.[32]

Partly because of these restrictions, among other reasons, the ability to access VR is not the same throughout the world. Besides more obvious barriers like affordability and access to the Internet, the components used for VR are restricted from being shipped to some countries. While there may be legitimate reasons for preventing this technology from being accessible to certain regions so that they cannot be repurposed for harm, this policy results in preventing people in those territories from accessing immersive environments? Other reasons, such as market viability or local restrictions, may convince manufacturers to not sell in certain countries. Meta, for example, lists only 23 countries, mostly in North America and Europe, where it will ship purchases of its Quest headset.[33] This may be a reasonable decision while VR remains a niche technology primarily used for entertainment, but as it becomes more widespread, or even necessary, part of everyday life, these restrictions will place more of a burden on providing access to the metaverse.

In addition to the problem of the digital divide, what other obligations do individuals, governments, and private companies have to provide access to the technologies that make access to immersive environments possible? As long as VR remains largely a medium of entertainment, it likely will be continued to be viewed as a superfluous product, not an essential need for people to become fully participant as citizens in their community. When VR transitions from an ephemeral activity for amusement to an indispensable need for communication, connecting with friends and loved ones, and building a career or running a business, what rights will people demand to access this technology? Will those already primed by VR games be favored (and rewarded) for performing work in the medium, and will these people be limited to regions where VR technology is available to them? If or when immersive technologies become a dominant form of interaction, it will be important to consider how to fulfill the obligation to provide VR technology to everyone who chooses to participate in the virtual world. Not doing so likely will lead to the solidification of economic, social, and class differences already present in society.

ACCESS TO VR SPACES

Ensuring equal access to VR technology is a necessary step to participating in virtual environments, but the ability to interact with other people and information contained within the metaverse is dependent on access granted to virtual spaces. In a virtual world, it is possible to design an environment in which anyone would be able to access any space they wanted to enter. One advantage of VR spaces over the real world is that there is no limitation of physical space. In the real world, location is everything: the physical laws of the universe prevent two people or objects from occupying the same space, and proximity to desirable spaces is valuable because of the limits of transportation. As of this writing, no one has developed the technology to instantly teleport a person from one physical space to another. Value is attached to desirable spaces, making, for instance, a densely populated waterfront property more valuable than the same amount of space on rural grassland far from other people.

A virtual world can mimic the real world, or it can invent its own rules. Designers can build immersive spaces that limit transport to the speed of human transportation, or they could have users instantly teleport to any space in the world. Likewise, if an exclusive club fills with avatars of users, an identical club in the visually same location can instantly open to accommodate more users. VR is limited only by the local spherical projection centered on the perspective of the user. How developers build spaces for social VR will determine how exclusive those spaces become. While the metaverse is limited only to the computing power required to store and render it, scarcity can be created artificially. Like the production of limited-edition collectible items or art prints, value can be manipulated by adjusting the relative availability and scarcity of a product or service.

To be sure, though, there are other limits to virtual spaces beyond computing power. In virtual worlds, scarcity of space is not determined by the physical areas in which users can occupy, but by the proximity of users to the other avatars with whom they seek to interact. Of course, a user's avatar could be duplicated across several virtual spaces simultaneously, but the user operating that avatar can perceive only one of those spaces at a time on a practical level and interact with other avatars in that one space. AI-controlled virtual avatars, or non-player characters (NPCs), can be set up to be in any, or every, space, but the exclusivity of real people controlling a singular virtual representation each creates an inherent scarcity that places a socially based limit on the possibilities for virtual interactions. A VR avatar can instantaneously appear in any virtual space, but the person controlling it can "be" in only one space at a time.

The metaverse that is described in *Snow Crash* contains two different types of scarcity: one imposed by the virtual real estate developer Global Multimedia Protocol Group in its design of a highway circumventing a virtual planet measuring a little over 65,000 kilometers in circumference, and one represented by the exclusive club that the main character Hiro can enter while others are left outside standing in the virtual rope line. The contrast between this virtual club and the shipping container where Hiro lives and accesses the metaverse demonstrates both the escape that a virtual world promises and the reproduction of limited resources in the virtual world.[34]

In its version of the metaverse, Meta Horizon Worlds places each user, as they enter the virtual environment, into their own personal space, where they can invite up to seven other members. Members can set their own rules of conduct in their personal space, including whether guests can use profanity, yell, or act in a bullying manner.[35] Users can venture from this space to other worlds with a click of a button, or they can hang out with their friends and watch a virtual movie. Though each room is limited in the number of people that can occupy it, members have created over 10,000 different worlds to visit.[36]

When spaces are limited to a few people, who should those people be? Meta implies that users likely will invite the friends they have or have made in the metaverse to share the same home space, launching together into a multiplayer game or remaining in the personal space to chat. Strangers, though, can also be invited, often to complete a squad for a game or to simply meet someone new online. Kashmir Hill's time spent in Meta's Horizon Worlds demonstrates that meeting strangers can alternate between both an enlightening and repulsive experience.[37] Sometimes the solution to the issue of misbehavior is to eject the offending player, but just as often a player's negative experience leads them to leave the space themselves from where the offending users have taken over. How much control should rest in the hands of a space's host, designer of the virtual space, or the guests requesting access to a space when determining which strangers are selected and allowed into a virtual room?

In some ways, virtual spaces are not that different from physical ones when considering their social makeup. Exclusive venues, like the annual gathering of corporate leaders and government policymakers at the Kansas City Federal Reserve Bank's economic symposium in Jackson Hole, Wyoming, generate international news for the people who show up there.[38]

Likewise, New York's Metropolitan Museum's annual fundraising gala, popularly known as the Met Gala, is advertised as "fashion's biggest night out" and generates its own red carpet coverage of the celebrities entering the venue.[39] Alternatively, other spaces work hard to drive people away or influence purchasing decisions, such as teen clothing stores playing hard rock and hip-hop music to convince parents to hand over their credit cards to their children rather than enter the store themselves, or convenience stores playing classical music to disperse the unhoused.[40] These and other psychological techniques can be used to influence people to view a space as inviting or inhospitable, and they are a softer way of controlling who enters the space.

The tension between control over access to virtual spaces and inclusivity is not limited to VR. Online symbolic spaces, from Internet chat rooms to online multiplayer video games and social media groups, have various rules to moderate the behavior of participants in these virtual spaces. Virtual spaces have established the norm of setting rules to regulate behavior in each space, and spaces function best when rules are explicit and clearly posted. While some spaces appoint a moderator, which usually is the creator of the space, to enforce these rules, other spaces, like the ones in Meta Horizon Worlds, allow for the polling of members in the space to vote out users who are perceived to be in violation of the rules.

Clay Shirky has written about the power of opening up virtual environments to harness the power of crowdsourcing. In *Here Comes Everybody: The Power of Organizing without Organizations*, Shirky argues that the barriers to accessing the power of organizing have been greatly reduced with the advantages of digital networking. The cost of organizing, from maintaining physical meeting spaces to physically traveling to meet, has been eliminated, making crowdsourced organizing online to address issues virtually free. Now people can use the Internet to share information, advocate for change, and find like-minded supporters (largely) without geographic limitations. The creation of spontaneous networks of social action, according to Shirky, disrupts traditional business and governing models not seen since the printing press. However, the benefits of online organizing are accompanied by the dangers of misinformation and the creation of groups that seek to do harm to others. In other words, the network makes no moral judgment on the groups that can organize with the diminished costs of digital networks.[41]

Many people have experienced the effects of disruptive behavior and malevolent organizing during the COVID-19 pandemic. As many people were sequestered in their homes, they used multiuser videoconferencing tools like Zoom or Google Meet to work, stay in touch with extended family, and catch up with friends. Online public gatherings that openly published links to join a space began to suffer from "Zoom bombing," where uninvited individuals attempted to disrupt the gathering with loud music, profanity, and hate speech, or by displaying obscene videos. Sometimes Zoom bombers would gain access to the virtual room's host controls, kicking members out, messaging other members, and effectively shutting down online events.[42] Organizations were forced to limit invitations to spaces, often sending out links only minutes before the beginning of an event to limit dissemination to Zoom bombers, while others limited who could join a session to invited guests.

The desire to moderate content to provide a civil experience online has itself led to criticism from those who argue that rules governing spaces are sometimes arbitrary or unfair to specific groups. For example, some conservative groups claim that moderation of online speech is a form of censorship that advances a political ideology. They argue that accusations of spreading conspiracy theories and other forms of disinformation are merely a political judgment about speech that does not conform to a narrow range of accepted thought. These groups accuse social media platforms of engaging in shadow banning, using an algorithm to hide or limit the sharing of posts without notifying users that their words and images are being curtailed.[43]

The balance between providing a free space open to everyone who can engage in a wide range of speech and debate while protecting against abuse and hate becomes even more delicate within immersive environments, where users are surrounded by the space they have entered. Unlike a framed screen, where users can look away, mute, or close a window, VR generates the illusion of copresence, where the actions of others more closely resemble physical proximity. If television is a window to the world, VR transports users to that world, heightening the impact of interactions within virtual spaces. Real-time interactivity also creates an environment more susceptible to harm. A study of video game players revealed that players experienced more engagement and arousal, including increased heart rate and pupil dilation, when they played a game online with other players than when they played alone offline.[44] Immersive environments also allow the communication of facial emotion cues and body language to add import to the language expressed in those spaces. This may encourage stricter rules for accessing immersive spaces than the ones currently employed for other online environments. However, exclusion from spaces often mirrors discrimination that is embedded in the analog world. Unwritten rules, social norms, and psychological nudges may reinforce the same disparities experienced in physical spaces, much like a visible security guard discourages members of oppressed groups from entering a physical space. To make immersive environments safer while remaining open, designers should be transparent about the rules of a space and make those rules clear to participants.

Access to VR content beyond multiuser environments is dependent on the availability of that content without restrictions or prohibitive costs. The model of VR as a gaming device restricts access to content, pricing software at rates comparable to video games available on PCs and gaming consoles. Changing this model to one in which producers think of VR as a portal to businesses, organizations, and community interest groups can expand access to those venues and appeal to more people. The European startup Atopia is working to expand access to cultural institutions like art institutions, museums, and performance spaces. Describing the company as "Netflix for culture," co-founder Anabell Vacano remarked, "Although over 50% of the EU population would travel mainly for cultural interest, many people are denied this opportunity, *e.g.* for health or financial reasons."[45] The application of VR as a portal to existing institutions may refocus efforts from selling hardware to empowering people to take advantage of cumulative knowledge and social connections.[46] Partnerships between VR developers, businesses, and nonprofit organizations can reduce the cost of offering services and open up access to more people who can make use of them.

When opening VR spaces to be accessible to more people, it will be important to address the incentives for joining a space. Researchers at the Brookings Institution have written that professors at minority-serving institutions are using VR for course instruction and recognize the need for students to develop skills that will become more in demand in their future careers, but many have expressed concern about promoting a new technology when many of the basic physical needs of so many of their students have yet to be met, including stable housing and food security.[47] Though VR can provide access to environments and simulations otherwise unavailable to students, the priority of physical stability and security cannot be ignored.

ACCESSIBLE VR

Even if all people were able to afford to purchase VR hardware and connectivity to participate in virtual spaces, not all bodies can adapt to the technology to engage in an immersive experience. Expecting someone to don an HMD and participate in VR experiences assumes that that person is capable of stereoscopic vision, binaural hearing, and a functioning vestibular system. Even with the physical capability of experiencing VR through a headset, many people become fatigued quickly within immersive environments and require breaks to suppress nausea induced by conflicts between the three-dimensional visual perception of accommodation, or changing focus between near and far objects, and vergence, the action of eye muscles to turn eyes inward to focus on near objects. In VR, near and far objects are both displayed on a screen that sits a fixed distance from a user's eyes, causing the conflict between vergence and accommodation.[48] Researchers recommend frequent eye exercises focusing on real-world objects to reduce fatigue and nausea, which limits the amount of continuous time one can spend in virtual worlds.[49]

Other digital technologies have addressed issues of accessibility through industry agreements and government regulation. The U.S. Department of Justice has taken the position that the Americans with Disabilities Act applies to websites created by businesses and federal, state, and local governments. Text descriptions of images, captioning of videos, and alternative ways of navigating web links, among other considerations, are all required for accessible web design.[50] Computer software companies have created accessibility tools within their operating systems and applications. For example, Microsoft allows users to set display and sound options for vision, hearing, neurodiversity, and mobility accommodations.[51] Tools like voice and eye navigation, adaptive joysticks, and haptic feedback devices allow users to interact with computers using the senses and motor actions available to them.

Instead of merely serving as an obstacle to users with disabilities, VR can serve as a tool to assist communities in meeting and accessing virtual worlds. The VRChat community Helping Hands invites deaf and hearing VR users together for the purpose of teaching and communicating in ASL and Dutch sign languages. Users can develop unique name signs that fit their personalities rather than having to sign each letter of their name, and members can click on words that link to videos showing someone signing that word.[52] One early interactive 360-degree video, Notes on Blindness, used animation paired with audio recordings of John Hull describing his experiences as he gradually lost his sight. The VR experience simulates for users what it is like to sense places, objects, and people using the character and direction of sound while listening to Hull's description of the psychological effects of going blind.[53] Researchers at Microsoft have developed a haptic cane controller so that blind users can navigate virtual worlds without sight, providing an alternative way of navigating immersive environments without a head-mounted display.[54]

To increase the accessibility of VR, Meta Platforms encourages developers to produce environments that can be experienced by a wider range of users. Its developer site states:

> Designing accessible applications means enabling people with varying levels of ability, vision, hearing, mobility, perception, cognition, or more, to interact with your content. Essentially, people should have the opportunity to interact with the experience and complete it as intended.[55]

Hand tracking, haptic feedback, captioning, user control over visual display adjustments, and accessible audio mixes are all considerations for accessible VR. An ethical code of VR for developers should include not only enabling access to these environments but also representing the spectra of abilities and communities within those environments. Avatars in virtual spaces need to represent different body types, including avatars using wheelchairs and other assistive devices, so that users may communicate to others how they are accessing immersive spaces if they so choose.

FREEDOM FROM VR

While most discussions about equitable implementation of VR have focused on ensuring people's ability of people to access it, often left unasked is the question of whether there should be a human right to refrain from using it at all. This may seem like a silly question to pose while VR is a niche product marketed mostly as a video game device, but it is important to set rules about limiting compulsory use before VR and AR become ubiquitous technologies. As VR progresses toward universal participation, the network effect predicts that the value of using VR will increase exponentially with the number of nodes in the network. Sharing information within immersive computing will become more efficient than through windowed browsers or offline methods, placing economic pressure on governments and organizations to make it a primary interface for conducting affairs. While policymakers work to ensure access to VR, what power will people have to exercise their right to freedom from using VR?

Beyond institutional policies and laws governing access to the technology, more subtle social and economic pressures may make it harder to avoid VR. For example, visiting family members who live far away in person may be a more meaningful experience than a phone call, but the time and cost savings of participating in a videoconference call create an incentive to use the technology over planning a trip and likely encourage more frequent contact with family. VR visits can become an even better technical solution, especially as avatars become photorealistic and generate the illusion of copresence. As VR technology progresses and becomes more widely adopted, will there be less incentive to make plans for travel, making it more difficult for people who do not use VR to engage in travel?

Even before the pandemic, videoconferencing was becoming a way to save costs for businesses. A 2019 Lifesize report on the technology showed that 47% of businesses surveyed had reduced travel by using videoconferencing, 51% expected remote work to become a component of their business, and the same percentage viewed businesses that used the technology as more innovative.[56] Post-pandemic, business leaders noted that the number of employees who work in a hybrid or fully remote mode has increased and will continue to increase over the next five years.[57] In this future environment, the cost of offline participation becomes even more prohibitive for more people. If access to information is a human right, the ability for everyone to participate in the access to that information independent of the technology to access it becomes even more important.

ETHICALLY ACCESSING VR

While VR is not as pervasive a technology as videoconferencing, social media, or Internet browsing, the immersive medium will improve over time and become used more as a substitute for in-person communication. A utilitarian model of ethics would require that VR and its underlying technologies would be provided to the most number of people as possible to access the network of immersive environments where more functions of everyday life will occur, much like the movement to reduce the impact of the digital divide has helped more people access global conversations and commerce through the Internet. However, while providing access to more people may help them, it may also harm others through the network effect by encouraging essential services to be relocated in VR environments. Such a future would create more obstacles for those who find it more difficult to adapt to the technology and those who just prefer to opt out.

A duty of care ethics may be a better framework for addressing questions of access to VR. This ethical framework considers the relationships between policymakers, developers, and consumers of VR. It is a model that incorporates the social connections between people and organizations to promote the well-being of all involved while considering the conflicting motivations and outcomes of setting policy. When examining the regulation of online content across several countries, Caio C. V. Machado and Thaís Helena Aguiar have found that content creators are being held more responsible for removing misinformation and hate speech from their platforms, but the use of automated tools to do so can lead to censorship of legitimate speech.[58] Building sustainable platforms that distribute responsibility between developers and users to ensure access while maintaining relationships of trust, while challenging, is a more just goal that uses a flexible approach to resolving tensions between freedom and security.

NOTES

1. Beilinson, Jerry. "Palmer Luckey and the Virtual Reality Resurrection." *Popular Mechanics*, May 28, 2014. https://www.popularmechanics.com/technology/gadgets/a12956/palmer-luckey-and-the-virtual-reality-resurrection-16834760/.

2. Kickstarter. "Oculus Rift: Step into the Game." January 30, 2016. https://www.kickstarter.com/projects/1523379957/oculus-rift-step-into-the-game.

3. Kosoff, Maya. "'The Final Medium Is Virtual Reality': Palmer Luckey, the Black Sheep of Silicon Valley, Opens Up About V.R., Zuckerberg, and Why He Supports Trump." *Vanity Fair*, October 10, 2018. https://www.vanityfair.com/news/2018/10/the-final-medium-is-virtual-reality-palmer-luckey-the-black-sheep-of-silicon-valley-opens-up-about-vr-zuckerberg-and-why-he-supports-trump.

4. King, Hope. "Oculus Founder Rails against Critics on Reddit." *CNNMoney*, April 26, 2016. https://money.cnn.com/2016/04/26/technology/oculus-palmer-luckey/index.html.

5. Apple Inc. "Introducing Apple Vision Pro: Apple's First Spatial Computer." *Apple Newsroom*. https://www.apple.com/newsroom/2023/06/introducing-apple-vision-pro/.

6. International Telecommunication Union. "Facts and Figures 2022: Latest on Global Connectivity amid Economic Downturn." *ITU Hub*, November 30, 2022. https://www.itu.int/hub/2022/11/facts-and-figures-2022-global-connectivity-statistics/.

7. Meta Platforms, Inc. "Meta Quest Account Payment Options." Accessed August 18, 2023. https://www.meta.com/help/quest/articles/accounts/payment-settings/meta-quest-payment-methods/.

8. PayPal. "Opening a PayPal Account." *PayPal User Agreement*, May 17, 2023. https://www.paypal.com/legalhub/useragreement-full.

9. World Bank. "Gains in Financial Inclusion, Gains for a Sustainable World." May 18, 2018. https://www.worldbank.org/en/news/immersive-story/2018/05/18/gains-in-financial-inclusion-gains-for-a-sustainable-world.

10. Federal Deposit Insurance Corporation. "2021 FDIC National Survey of Unbanked and Underbanked Households." July 24, 2023. https://www.fdic.gov/analysis/household-survey/index.html.

11. U.S. Congress. Telecommunications Act of 1996, Pub. L. No. 104, § 706, 104 (1996). https://www.congress.gov/104/plaws/publ104/PLAW-104publ104.pdf.

12. Lohr, Steve. "A Nation Ponders Its Growing Digital Divide." *The New York Times*, October 21, 1996. https://www.nytimes.com/1996/10/21/business/a-nation-ponders-its-growing-digital-divide.html.

13. Cooper, Mark, and Gene Kimmelman. "The Digital Divide Confronts The Telecommunications Act Of 1996: Economic Reality Versus Public Policy." *Consumer Reports*, September 14, 1999. https://advocacy.consumerreports.org/wp-content/uploads/2013/04/telecom1-0299.pdf.

14. Norris, Pippa. *Digital Divide: Civic Engagement, Information Poverty, and the Internet Worldwide*. Cambridge University Press, 2001. https://doi.org/10.1017/CBO9781139164887.

15. Vogels, Emily A. "The Longer and More Often People Use Facebook, the More Ad Preferences the Site Lists about Them." *Pew Research Center*, December 3, 2019. https://www.pewresearch.org/short-reads/2019/12/03/facebook-ad-preferences-linked-to-frequency-of-use-age-of-account/.

16. Vassilakopoulou, Polyxeni, and Eli Hustad. "Bridging Digital Divides: A Literature Review and Research Agenda for Information Systems Research." *Information Systems Frontiers* 25, no. 3 (June 2023): 955–969. https://doi.org/10.1007/s10796-020-10096-3.

17. O'Neil, Cathy. *Weapons of Math Destruction: How Big Data Increases Inequality and Threatens Democracy*. New York: Crown, 2016.

18. Meta Platforms, Inc. "Oculus Quest 2 Store: VR Games, Apps, & More." *Oculus*. Accessed August 24, 2023. https://www.oculus.com/experiences/quest/.

19. Cline, Ernest. *Ready Player One*. New York: Crown Publishers, 2011.

20. Stephenson, Neal. *Snow Crash*. Bantam Spectra Book. New York: Bantam Books, 2008.

21. Pimentel, Daniel, Maxwell Foxman, Donna Z. Davis, and David M. Markowitz. "Virtually Real, But Not Quite There: Social and Economic Barriers to Meeting Virtual Reality's True Potential for Mental Health." *Frontiers in Virtual Reality* 2 (February 17, 2021): 627059. https://doi.org/10.3389/frvir.2021.627059.

22. Ma, Wayne. "Apple Hopes Chinese Suppliers Can Solve Vision Pro's Display Challenges." *The Information*, August 29, 2023. https://www.theinformation.com/articles/apple-hopes-chinese-suppliers-can-solve-vision-pros-display-challenges.

23. Shah, Agam. "Intel's Lack of Potent Graphics May Hurt in VR and AI Chase." *PC World*, August 18, 2016. https://www.pcworld.com/article/416087/lack-of-a-high-quality-gpu-may-hold-intel-back-in-its-ai-and-vr-chase.html.

24. OpenAI. "Introducing ChatGPT." *OpenAI*, November 30, 2022. https://openai.com/blog/chatgpt.

25. McKinsey & Company. "The State of AI in 2023: Generative AI's Breakout Year." *McKinsey*, August 1, 2023. https://www.mckinsey.com/capabilities/quantumblack/our-insights/the-state-of-ai-in-2023-generative-AIs-breakout-year.

26. Intel. "What Is a GPU? Graphics Processing Units Defined." Accessed September 1, 2023. https://www.intel.com/content/www/us/en/products/docs/processors/what-is-a-gpu.html.

27. Reznik, Audrey, Troy Nelson, Kaitlyn Abdo, and Christina Xu. "Why GPUs Are Essential for AI and High-Performance Computing." *Red Hat Developer*, November 21, 2022. https://developers.redhat.com/articles/2022/11/21/why-gpus-are-essential-computing.

28. Dave, Paresh. "Nvidia Chip Shortages Leave AI Startups Scrambling for Computing Power." *Wired*, August 24, 2023. https://www.wired.com/story/nvidia-chip-shortages-leave-ai-startups-scrambling-for-computing-power/.

29. Lee, Joyce, and Heekyong Yang. "Nvidia Supplier SK Hynix Says HBM Chips Almost Sold out for 2025." *Reuters*, May 2, 2024. https://www.reuters.com/technology/nvidia-supplier-sk-hynix-says-hbm-chips-almost-sold-out-2025-2024-05-02/.

30. Leswing, Kif. "Nvidia Stock Falls after U.S. Government Restricts Chip Sales to China." *CNBC*, August 31, 2022. https://www.cnbc.com/2022/08/31/nvidia-stock-falls-after-us-government-restricts-chip-sales-to-china.html.

31. Kharpal, Arjun. "Nvidia Warns More Semiconductor Curbs Will End U.S. Chipmakers' Ability to Compete in China." *CNBC*, August 24, 2023. https://www.cnbc.com/2023/08/24/nvidia-says-ai-chip-export-curbs-to-china-will-hit-us-chipmakers.html.

32. Middleton, Joe. "US Restricts Exports of Nvidia AI Chips to Middle East." *The Guardian*, August 31, 2023. https://www.theguardian.com/technology/2023/aug/31/us-restricts-exports-of-nvidia-ai-chips-to-middle-east.

33. Meta Platforms, Inc. "Meta Quest Supported Countries and Regions." *Meta Store*. Accessed May 13, 2024. https://www.meta.com/help/orders-and-returns/articles/quest-supported-countries/.

34. Stephenson, *Snow Crash*.

35. Meta Platforms, Inc. "Personal Space in Meta Horizon Worlds." Accessed September 1, 2023. https://www.meta.com/en-gb/help/quest/articles/horizon/explore-horizon-worlds/personal-space-in-meta-horizon-worlds/.

36. Meta Platforms, Inc. "Making VR More Social with Meta Horizon Home." *Meta* (blog), June 14, 2022. https://about.fb.com/news/2022/06/making-vr-more-social-with-meta-horizon-home/.

37. Hill, Kashmir. "This Is Life in the Metaverse." *The New York Times*, October 7, 2022. https://www.nytimes.com/2022/10/07/technology/metaverse-facebook-horizon-worlds.html.

38. Irwin, Neil, and Courtenay Brown. "A Beginner's Guide to the Fed's Big Conference in Jackson Hole." *Axios*, August 25, 2022. https://www.axios.com/2022/08/25/jackson-hole-fed-guide.

39. Nast, Condé. "Met Gala 2023: Celebrities, Red Carpet, Theme & More." *Vogue*. Accessed September 1, 2023. https://www.vogue.com/tag/event/met-gala.

40. Waude, Adam. "Ambient Influence: How Background Music In Stores And Supermarkets Influences Consumers' Purchasing Decisions." February 23, 2016. https://www.psychologistworld.com/behavior/ambient-music-retail-psychological-arousal-customers.

41. Shirky, Clay. *Here Comes Everybody: The Power of Organizing without Organizations*. New York: Penguin Press, 2008.

42. Claire Seungeun Lee has documented the prevalence of cyber hate speech in the United States, including expressions of racism against Asian Americans, African Americans, and Jewish Americans in cases of Zoom bombing during the pandemic: Lee, Claire Seungeun. "Analyzing Zoombombing as a New Communication Tool of Cyberhate in the COVID-19 Era." *Online Information Review* 46, no. 1 (January 25, 2022): 147–163. https://doi.org/10.1108/OIR-05-2020-0203.

43. Levitz, Eric. "The 'Twitter Files' Is What It Claims to Expose." *Intelligencer*, December 10, 2022. https://nymag.com/intelligencer/2022/12/twitter-files-explained-elon-musk-taibbi-weiss-hunter-biden-laptop.html.

44. Juvrud, Joshua, Gabriel Ansgariusson, Patrik Selleby, and Magnus Johansson. "Game or Watch: The Effect of Interactivity on Arousal and Engagement in Video Game Media." *IEEE Transactions on Games* 14, no. 2 (June 2022): 308–317. https://doi.org/10.1109/TG.2021.3073084.

45. De Marzo, Stefano. "Munich-Based Atopia Secures Six-Figure Pre-Seed to up the Cultural Industry Game with VR Technologies." *EU-Startups*, August 30, 2023. https://www.eu-startups.com/2023/08/munich-based-atopia-secures-six-figure-pre-seed-to-up-the-cultural-industry-game-with-vr-technologies/.

46. Park, Sang-Min, and Young-Gab Kim. "A Metaverse: Taxonomy, Components, Applications, and Open Challenges." *IEEE Access* 10 (2022): 4209–4251. https://doi.org/10.1109/ACCESS.2021.3140175.

47. Lee, Nicol Turner, Rashawn Ray, Samantha Lai, and Brooke Tanner. "Ensuring Equitable Access to AR/VR in Higher Education." *Brookings*, September 6, 2022. https://www.brookings.edu/articles/ensuring-equitable-access-to-ar-vr-in-higher-education/.

48. Zhang, Sarah. "The Obscure Neuroscience Problem That's Plaguing VR." *Wired*, August 11, 2015. https://www.wired.com/2015/08/obscure-neuroscience-problem-thats-plaguing-vr/.

49. Thai, Kien T. P., Sungchul Jung, and Robert W. Lindeman. "On the Use of 'Active Breaks' to Perform Eye Exercises for More Comfortable VR Experiences." In *2020 IEEE Conference on Virtual Reality and 3D User Interfaces Abstracts and Workshops (VRW)*, 468–476. Atlanta, GA, USA: IEEE, 2020. https://doi.org/10.1109/VRW50115.2020.00099.

50. ADA.gov. "Guidance on Web Accessibility and the ADA," March 18, 2022. https://www.ada.gov/resources/web-guidance/.

51. Microsoft. "Accessibility Technology & Tools." *Microsoft Accessibility*. Accessed September 3, 2023. https://www.microsoft.com/en-us/accessibility.

52. Bayliss, Ben. "Deaf Players Create VRChat Community and Teach Sign Language." *DualShockers*, November 18, 2019. https://www.dualshockers.com/vrchat-deaf-community-teach-sign-language/.

53. Carvajal, Andrea. "Notes on Blindness Review." *VR Voice* (blog), August 31, 2019. https://vrvoice.co/notes-on-blindness-review/.

54. Hagen, Alexis. "VR without Vision: Immersive Sensory System for People Who Are Blind." *Microsoft Research* (blog), April 7, 2020. https://www.microsoft.com/en-us/research/blog/bringing-virtual-reality-to-people-who-are-blind-with-an-immersive-sensory-based-system/.

55. Meta Platforms, Inc. "VR Accessibility Design: Controller Mapping, Input, and Feedback." *Oculus Developers*. Accessed September 3, 2023. https://developer.oculus.com/resources/design-accessible-vr-controls/.

56. Lifesize. "Lifesize 2019 Impact of Video Conferencing Report: More Collaborative Workplace Cultures Have Led to Spike in Video Communication," August 27, 2019. https://www.lifesize.com/press/impact-video-conferencing-report/.

57. Bloom, Nicholas, Jose Maria Barrero, Steven Davis, Brent Meyer, and Emil Mihaylov. "Survey: Remote Work Isn't Going Away—and Executives Know It." *Harvard Business Review*, August 28, 2023. https://hbr.org/2023/08/survey-remote-work-isnt-going-away-and-executives-know-it.

58. Machado, Caio C. V., and Thaís Helena Aguiar. "Emerging Regulations on Content Moderation and Misinformation Policies of Online Media Platforms: Accommodating the Duty of Care into Intermediary Liability Models." *Business and Human Rights Journal* 8, no. 2 (June 2023): 244–251. https://doi.org/10.1017/bhj.2023.25.

Chapter 5

Privacy in VR

When users put on a VR headset, they can direct their gaze to any point within their view. If an object or avatar is out of their view, they simply turn their head to see it. As long as it is not blocked by other virtual objects closer to the user, anything can be viewed by those using VR; users control what they see by simply turning and looking at what they want. The powerful feeling of control is greater than with older media like film and television, where a director determines what audiences see and for how long. A user often is alone in a room while wearing their headset; they control which spaces they visit and when, which games to play, and which other users to interact with, all within the privacy of the room they occupy, often within their private home. That power is illusory, however, because while the user is experiencing an immersive environment in a VR device, that device is gathering information about the user.

What information VR collects, where that information goes, and who has access to that information all relate to ethical questions about the rights to privacy when engaging with VR. While VR is by no means the first digital technology for which critics have expressed concerns about privacy and data collection, the immersive nature of VR gathers much more personal information than other media, including biological and physiological data that can reveal more about a user than they realize. Beyond what content users choose to experience, their interactions within a specific experience generate data that is gathered by the device, including movement, eye gaze, pupil dilation, pulse, and reflex time. Not only can the technology record the choices users make in VR, but it also possesses the ability to record the physical and emotional states of users when they engage with the technology. A comprehensive ethics of VR must include the role of data collection and sharing when experiencing immersive environments.

DATA PRIVACY AND DIGITAL TECHNOLOGIES

Data privacy has become a primary issue of concern for many people, especially since the reporting of social media data collection and dissemination. In 2014, the political communications firm Cambridge Analytica gained access to private information from approximately 87 million Facebook users to build political and psychographic profiles about them. Most of these people never consented to sharing their data, but because they were connected to Facebook friends who had downloaded a digital Facebook app developed by Aleksandr Kogan called "thisisyourdigitallife," their profile information, likes, and other information was collected to build personality profiles. *The New York Times* reported in 2018 that Cambridge Analytica used this information to target political ads, both to drive turnout for Donald Trump and suppress the vote for Hillary Clinton, both of whom were running for president of the United States.[1]

The revelation launched investigations in both the United States and United Kingdom, where Cambridge Analytica was also was accused of interfering in the vote to determine whether the United Kingdom should exit the European Union. Whistleblower Christopher Wylie described the use of people's Facebook data as building "models to exploit what we knew about them and target their inner demons."[2] Brittany Kaiser, another whistleblower who worked at Cambridge Analytica, provided documents to authorities that showed how the company developed its strategy first in African countries, where privacy laws are less stringent, before she sold to clients "a political tool that was ruthless, effective, and most dangerous of all for democracy, scalable."[3]

Soon after the scandal was made public, vice president and deputy general counsel of Facebook (now Meta), Paul Grewal, asserted in a blog post that the company has strengthened its data privacy protections, including intensifying its app review process and providing users with more control over permission to share their data.[4] Despite these assurances, a hacker forum posted the personal information of over 530 million Facebook users in 2021, though Facebook asserts the data was obtained before the company strengthened its data protections.[5] The Federal Trade Commission fined Facebook $5 billion in 2019 for its role in violating the privacy of its users.[6] However, according to researcher Margaret Hu, the settlement with the FTC still did not adequately address what she labels "black boxing the voter," or using nontransparent digital code and data to assert authority over or influence the decisions of groups of people without their knowledge.[7]

In 2022, the problem of the black box resurfaced with the release of ChatGPT, an AI tool that generates text and code in response to prompts typed in a text window. The research lab that produced ChatGPT, OpenAI, states that they released this chatbot to the public to allow everyone to access the power of AI, instead of the previously dominant model of corporations and governments building and using closed AI systems for their own benefit. "Generally speaking, we think more usage of AI in the world will lead to good," OpenAI co-founder Sam Altman wrote in the organization's blog. "We believe that democratized access will also lead to more and better research, decentralized power, more benefits, and a broader set of people contributing new ideas."[8] ChatGPT is free to use for anyone who signs up for an account, though access to its more advanced model costs $20 per month.

This large language model, like most AI, is trained on a large amount of data, and its learning is reinforced by the feedback of users, which in turn becomes a part of its dataset. The black box problem with AI is the opaque nature of its responses. While an AI model is capable of generating responses to prompts, users (and even its programmers) do not know the process or data used in generating those responses. AI suffers from a lack of accountability that reduces the amount of trust users can place in its responses. For example, AI large language models sometimes produce false sources and misinformation that observers have called "hallucinations." OpenAI researcher Lilian Weng stated:

> Current deep learning models are not perfect. They are trained with a gigantic amount of data created by humans (e.g., on the Internet, curated literature) and unavoidably absorb a lot of flaws and biases that long exist in our society. For example, when DALL·E was asked to portray a nurse, it would only generate female characters, or for a professor, it would only generate white people. The model captures biases in real world statistics or biases in our training data.[9]

Weng complimented OpenAI for taking this problem seriously by adding "human-in-the-loop evaluation," but the problem, as other AI researchers have suggested, may be an inherent part of AI, necessitating that humans check the output these algorithms generate.[10]

Businesses that are concerned about revealing proprietary data to ChatGPT can take advantage of privacy protections, as OpenAI has announced a special enterprise system that allows subscribers to train a closed AI model with their data without that data being used to train the version available to the public.[11] This service is available at a cost, which has contributed to OpenAI making $80 million in revenue every month and being on track to make $1 billion annually.[12] Its revenue is expected to increase even more after Apple announced that it plans to integrate ChatGPT features into its iOS, including within iPhone and iPad devices.[13] Those who use the free service continue to have the prompts they supply to ChatGPT incorporated into the AI black box, without knowing if or how that data is used.

VR may help AI models by incorporating data about how humans react to digital content and passthrough video of the real world. Already, Alphabet's Google is using data from YouTube videos to train its AI model, which the company claims is five times more powerful than ChatGPT-4. The concern that AI will diminish in quality as it trains on AI-generated content posted on the World Wide Web has led developers to look for new sources of human-created content, including music and videos. Major music labels are negotiating with YouTube to prevent their artists from being used in training data, but everyone else who posts videos to the site can expect that their content will become part of the AI model.[14] VR can provide an excess of human-generated data which likely will be used to train AI algorithms to guide NPCs and other computer-controlled activity in the metaverse. ChatGPT already is available on Apple Vision Pro and likely will become integrated into other VR devices.[15]

The documentary *The Social Dilemma* illustrates the effect of gathering information from social media users in order to sell that information to advertisers and other interested parties. Online platforms serve content to users to engage them for longer periods of time, often at the expense of their mental health, while gathering information about their online activity and matching it with other identifiable information to sell to advertisers and anyone else willing to pay. Tristan Harris, co-founder of the Center for Humane Technology, states plainly in the documentary: "If you're not paying for the product, you are the product."[16] With social media platforms, engagement is measured in clicks, but VR can provide more personal and identifiable data, including eye gaze, pace of movement, and emotional states. If social media moves into the metaverse and offers virtual spaces for free to users, it is likely this business model will be extended into immersive spaces so that the companies can gather and sell data about the users engaged in these spaces. These companies also will be incentivized to keep users in these spaces for longer periods of time so that they can gather more information and serve engaged audiences to advertisers.

THE VALUE OF PERSONAL DATA

Businesses likely would not be interested in the mass collection of personal data if there were no profit incentive for doing so. The commodification of information is driven by a demand to gain insights into the habits, purchasing decisions, voting preferences, emotional predispositions, and other behaviors of a large number of people. This data is made useful because digital technologies have made it cost-effective to collect billions of data points and process those bits of data to glean useful information from them. The collecting and selling of information certainly existed in the pre-digital age: historian

William Leach has written that, with the rise of consumer products in the late nineteenth century, manufacturers sought to understand the markets to which they wanted to sell mass-produced products and consumable goods. They created the idea of a mass-produced luxury, a democratic distribution of opulence, available for a price. Leach argues that businesses "turned rapidly to new methods of marketing and to the dissemination of strategies of enticement."[17] This required an understanding of who made up the market and how it could be persuaded to participate in a consumer society. The coordinated marketing of consumer goods required centralization and capital investment, limiting the power of data collection to large companies and organizations.

With the rise of mass media, understanding public opinion and predicting behavior became big business. *The Literary Digest* collected these opinions by having readers mail back postcards with their responses to questions. However, journalism professor George Gallup developed a more scientific means to poll representative samples of people to measure opinion more accurately.[18] The use of telephone polling made this process more efficient, and the emergence of computers as data processing devices provided companies with a means for collecting and storing data about people. While pollsters are often associated with political campaigns and governments, the much larger industry of market research, valued at over $81 billion, collects much more personal information to build consumer profiles in order to sell products and services to consumers.[19]

As more people use debit cards, credit cards, and other forms of digital payments to make purchases, that data has become a profit center for banks. Disclosed in mandated privacy policies, many banks share information with "nonaffiliates," meaning, according to *Washington Post* technology columnist Geoffrey A. Fowler, anyone who is willing to pay for this information. In addition, large credit card companies like Mastercard and Visa sell aggregated data in groups as small as 50 people, which other marketers can use to connect with web traffic data and online purchases in order to identify individual consumers.[20]

The amount of data available from several sources has led to the emergence of data brokers, companies that organize data to make actionable recommendations to clients. One of the largest companies involved in this practice is Acxiom, a division of Interpublic Group. Among the other services that it offers, Acxiom promotes its plan for companies to participate in the "new data economy" by connecting the data of each client to benefit their marketing efforts or generate their own data about their customers that they can sell to others. It assists companies in building what it calls a "customer intelligence strategy" to help companies "find and retain customers."[21] A case study that Acxiom presents shows how restaurateurs located in business districts can boost sales by using their point-of-sale data combined with "diner recognition" data that includes lifestyle changes and behavioral information to target those customers with offers that would make them more likely to make a purchase.[22] Another data broker, Epsilon Data Management, claims it holds the largest customer database, which it has built from public records, customer-reported surveys, magazine publishers, product warranty card returns, and purchasing data, among other sources.[23] Left unsaid is the ability of companies, using this data, to employ psychological techniques to nudge people to make purchases that they normally would not have made because these companies possess identifiable information about customers' habits, compulsions, and other psychological dispositions.

While social media companies like Meta have been the focus of scrutiny about the data they gather about its users so that they can sell targeted advertising, the selling of online data to brokers for aggregation stretches far beyond the largest tech companies. Email campaigns, websites, affiliate programs, and mobile apps all have the ability to track visitors, and many do so to generate revenue. Many people already are familiar with cookies, or data stored in a user's web browser when they visit a website. Cookies can store a user's browsing history and provide useful functionality for configuring pages when users visit again, but third-party cookies allow advertisers to track users across several sites and store information that includes search queries, products viewed but not purchased, and time spent on each site.[24]

Even if users delete cookies from their browser or use a plug-in to block cookies, data gatherers have many other ways of collecting information about people's online activity. A web beacon hides code within a page to send information about the IP address visiting the site, what operating system and device a person is using, where they are located, whether an email was opened or not, or if a user interacted with an ad. Similarly, a tracking pixel is an imperceptible graphic that essentially tags viewers and can follow them across their browsing activity.[25] Many users are unaware of the extent to which companies are able to track web activity. One study offered a web tracking identifier to over 400 participants and surveyed them about what they had discovered. While most users already were aware that some websites engage in tracking, they were surprised by the extent of the tracking and how much advertisers knew about them.[26] These methods of tracking users and providing data about them are invisible to users of digital technologies but provide valuable information to marketers and companies seeking to monetize this data.

PRIVACY AND IDENTITY

Despite many digital companies' claims that the data they collect are anonymized and used to track consumers' habits in aggregate, their real value is in the ability to correlate data points from several sources to discover the identity of each person through the multiple ways individuals represent themselves online. One service Acxiom offers is Identity Resolution, or the ability to identify individual customers by correlating aggregated data from multiple sources. Acxiom claims matching individual customers to their personal data gives companies a competitive advantage by allowing them to provide unique customer experiences or modify behavior for specific customers based on their purchasing power and other relevant data.[27]

Many companies use a version of digital fingerprinting to identify individual users. Multiple studies, according to *Wired*, show that fingerprinting, which consists of examining the language, browser, keyboard layout, operating system, and other technical data about a user, is unique in about 80 to 90% of individuals. About a quarter of the top 10,000 websites have digital fingerprinting code embedded in their pages, sometimes without the knowledge of the website owner.[28] Despite promises to the contrary, the value in identifying the activity and preferences of each person drives the motivation of companies to collect this information.

Though the Internet had originally provided some degree of anonymity for users, the ability to track activity and connect data points to reconstruct unified individual identities has allowed those able to pay for these services to lift the privacy veil. The reason why this matters is that the concern about privacy and digital technologies ultimately comes down to a question of power: the ability of companies to identify individual users and understand their preferences is a two-way mirror. First, many platforms allow users to opt out of sharing personal information, but the default condition is to share personal data, tilting the scales in favor of data collection. Second, discovering what data is being shared and with whom can be a labyrinthine affair, as this information is often buried within the legal language of pages-long terms and conditions statements. According to a 2019 survey conducted by the Pew Research Center, 74% of Americans reported that they never or only sometimes read the privacy policies of companies before they agreed to their terms and conditions. Of those who have read a privacy policy, only 22% say they read it all the way through. The same survey found that 79% of Americans do not feel confident that companies will "publicly admit mistakes and take responsibility when they misuse or compromise their users' personal data."[29]

Third, in many cases, the choice of whether or not to share personal information is not really a choice: some applications combine data collection with requirements for proper operation of the software. In many cases, personal data is needed to offer useful suggestions, store images, and unlock features otherwise unavailable. Natasha Lomas and Romain Dillet found that the terms and conditions statement allowing people to use Apple iTunes ran 20,000 words, while Facebook's statement was about 15,000 words.

Spotify's statement, according to the authors, "talked about collecting your photos, location, contacts and more."[30] The authors note that these statements are a form of legal language that developed in the pre-digital age and are often written to protect companies at the expense of the consumer. In order to use the Meta Quest, users must agree to Meta's Terms and Conditions statement, which includes an over 19,000-word privacy policy. The policy repeatedly claims that it is necessary to collect information about user identity to keep the platform safe for everyone and allows Meta to report abusive or otherwise illegal behavior to authorities. The data it collects and stores includes

> the time, frequency and duration of your activities on our Products . . . purchases . . . device characteristics and device software . . . location-related information . . . age, the gender you provide . . . messages you send and receive . . . content you provide . . . information from cookies and similar technologies . . . hashtags you use . . . what you're doing on your device, like whether our app is in the foreground or if your mouse is moving.

among many other data.[31] While Meta allows users to control some amount of data sharing, those settings are buried in a privacy menu three levels deep in a settings menu, making it unlikely for casual users to access these options. Though many of these data points can be combined to identity a user, some other shared data outright identifies users without a need to compare against other information. If a user chooses not to accept these terms and conditions, they are not able to access the platform.

While Meta's Terms and Conditions apply to all its products, including Facebook, Instagram, and WhatsApp, this same agreement takes on new significance when it is applied to its VR devices. The ability of sensors in a VR headset to gather data about human activity goes far beyond a traditional keyboard and mouse. Watching a 360-degree video in VR may feel like a solitary activity, but an advanced VR HMD has the capability of gathering information about the direction of viewing, gaze, reaction time, pupil dilation, and pulse, in addition to which videos a user chooses, how long they watch, when they pause or exit a video, and other information. The Apple Vision Pro headset is equipped with a LiDAR scanner and True Depth camera to model a user's living or workspace in three dimensions, and its Optic ID records biometric data of a user's iris to identify who is wearing the headset.

To be sure, much of this data is needed to allow the device to operate or to provide convenience for the user: the HMD must report where a user is looking to properly render the video to display. However, because these devices have the capability to gather and store this data, ethical rules are needed to guide what data may be released to the platform, other users, and third parties who are willing to pay for insights this data may provide. For example, a health insurance company may be interested in the biometric data that may reveal any medical issues of users, while an auto insurance company may be interested in the reaction times of users to predict the likelihood of future accidents. Marketers could infer from the size of a room paired with location data the income level of a user, and they can steer them to make purchases by displaying different objects and messages in virtual environments based on this information. The stories presented in *Snow Crash* and *Ready Player One* would likely take a different turn if the platform owners could detect that Hiro accessed the Metaverse from a storage container and Wade entered the OASIS from an abandoned junkyard van in the Stacks. Users should be aware of what personal and identifiable data can be stored and tracked in VR, and platform managers must make transparent what data is shared with whom and for what purposes.

Despite the efforts of manufacturers to assure users that their identity is protected by anonymizing data before they share it with partners and affiliates, a 2020 study of users watching 360-degree videos while wearing a VR headset found that, with as little as five minutes of tracking data that the HMD provided, researchers were able to identify 95% of viewers. The researchers concluded that nonverbal data can be just as revealing as written or spoken responses to prompts.[32] Users may be misled into believing that their identity is secured with promises of anonymity and tools like private browsing modes, when the tracking data alone from using a VR headset is enough to establish a user's identity. This concern is compounded by the fact that users may let their guard down while they interact within virtual environments because their senses are immersed in that environment and not in the physical world. While users watch the immersive environment around them, their headset is watching them.

SURVEILLANCE IN VR

The ability to connect a user's behavior with their identity certainly is useful to advertisers and companies who want to measure the patterns and habits of their customers, but the collection and dissemination of data that establishes identity in real time creates an even more existential issue for users: the ability for others to conduct surveillance of their activities. Harvard professor Shoshana Zuboff has written about the commodification of identifiable personal data without the express consent of the people producing that data. According to Zuboff, this system operates under the metaphor of the capitalism of physical goods, where the personal data of people is treated like a raw material that is to be extracted and refined to be sold at a profit. The relationship between the surveillance capitalists and individuals whose data is mined is asymmetrical, as the collected data can be used to influence behavior and control activity. Furthermore, the market for digital surveillance is largely invisible to those affected by it: people may not know why they were not called for a job interview, given less favorable terms for a loan, or stopped by police more often than others, when the processing of data collected and sold by surveillance capitalists played a part in these decisions, whatever the intent of those being surveilled.[33]

Zuboff argues that the only possibility for reining in the power imbalance created by surveillance capitalism is for governments and citizens to demand rules that limit what data can be collected and shared and for surveillance capitalists to seek explicit permission to use data that is inherently owned by each individual. A systematic transparency of data collection is needed, especially when that data is used to make policies and future decisions that govern people. Because surveillance capitalism has the power to identify individuals through their digital data, which allows for the remote surveillance of them, strict regulations that are democratically set should restrict this power so that it benefits the common good over special interests or the profit motive of those selling this information.

This becomes even more important for VR as immersive technologies become more entrenched as a form of mediation between people. Not only does VR have the capability to collect and transmit personal data, including individuals' physiological states, but it can also adjust virtual environments in real time to adapt to that data. Already, medical researchers have discovered therapeutic uses of VR for rehabilitation, but they argue that interactivity is key for its effective use, including the need for functional feedback in real time.[34] Listing effective traits of VR therapy, one group of researchers notes that:

> An adaptable VR task would be able to scale the difficulty of the task in real-time to adjust to the user's individual needs. Explicit feedback could be given through visual or auditory cues which allow the user to know if they are performing the task correctly, and task goals can be set at the beginning of the VR task. Interactivity speaks to the synchronicity of the user's behavior in the real-world to the outcome in the virtual world. A VR task with high interactivity could have an avatar, or similar representation of the user, which performs actions in the VR in concert with the user's movements in the real-world. Adding the option of competition to a VR would allow the user to perform the task in the VR while competing against a virtual opponent.[35]

This last suggestion, competition among virtual opponents, demonstrates the possibility of real-time interaction that generates data, which could be used to modify the responses of participants. Unmediated interpersonal interactions lead to socialization that ingrains rules about how to behave around others. However, VR-mediated interactions may be modified by designers to shape that socialization by changing the reactions of others in real time. As an extreme example, a smile by one VR participant may be represented as a frown by that person's avatar, or a computer-controlled avatar may modify behavior based on the input by coders of the experience. By gamifying experiences in VR, the modification of these interactions can be invisible to users, as they enjoy the experience of the virtual environment without being aware of its purpose.

VR is particularly effective for exposure therapy, where people who suffer from phobias can gradually be exposed to a stimulus to reduce that fear. One study found that people with acrophobia who used a VR app over two weeks reduced their fear of heights by about half.[36] Other groups of researchers have treated arachnophobia by placing patients in virtual environments where they can see spiders in different scenarios. The VR app Itsy has helped reduce the fear of spiders in as little as three hours,[37] and another study shows that exposure in VR is as effective as an in-person exposure session, though the effects diminish over time.[38] The fear of public performance is difficult to treat with exposure therapy because it requires an audience, but VR exposure can simulate large amphitheaters with a responsive audience.[39]

Creating enjoyable VR experiences certainly may be beneficial for therapeutic uses, where the length of exposure is correlated to the achievement of the therapy's desired outcome, but these same techniques could be used to conceal behavioral modifications made by marketers or political actors. A free VR gaming app, for instance, that uses the same techniques as VR exposure therapy apps could be produced to encourage players to habituate impulse purchasing or engender a fear of a specific political or ethnic group. Moreover, despite the professional ethics of clinicians, the physiological and psychological data that can be collected from the use of commercially available VR therapy apps easily could be co-opted by other parties. For example, a platform host or app creator could demand in its list of terms and conditions the need to transmit biometric and other personal data to a central server for collection and analysis for the purposes of technical troubleshooting and improving the software. However, once that data is in the hands of the designer, the enticement to monetize that data may overcome ethical concerns for the privacy of its users. In fact, if the app creator works for a publicly traded company, there may be a fiduciary responsibility to that company's shareholders to monetize that data as a profit center and to justify further investment.

The digital surveillance of VR users may be justified in cases where significant harm may be committed by them, but law enforcement historically has been restrained by legal restrictions when it comes to protecting the privacy of citizens. In the United States, the Constitution's Fourth Amendment right to privacy applies unless a search warrant is presented that establishes a probable cause for search or seizure, and the due process clause of the Fourteenth Amendment restricts government agencies from large-scale collection of personal data. However, the justifications of public safety and national security have motivated officials to engage in activities that violate the right to privacy and due process. In 2018, the Supreme Court ruled in *Carpenter v. United States* that police cannot search through historical location records of cellular phone towers without a warrant.[40] However, since that ruling, government agencies have found a way around this restriction by purchasing the same information from private data brokers. According to NYU School of Law professor Elizabeth Goitein, the law forbids government organizations from forcing companies to release private data without a warrant, but they can incentivize the practice simply by purchasing the data from any private source that has legally collected it.[41]

The problem with the mass collection of private data is that it is often used to justify policies or actions that are problematic. In *Weapons of Math Destruction*, Cathy O'Neil demonstrates how the algorithmic processing of personal data amplifies biases and systematic injustices while absolving policymakers of responsibility for the decisions they claim the algorithms have made.[42] For example, the identification of antisocial or potentially criminal behavior may be used by law enforcement to create a probability score of recidivism. Already, many law enforcement agencies use risk assessment algorithms to help determine sentencing after someone is convicted of a crime. However, a *ProPublica* report from 2016 revealed that the risk assessment algorithm used for people arrested in Broward County, Florida, was only "somewhat more accurate than a coin flip" and often was biased against Black defendants, falsely identifying them as future criminals at twice the rate of white offenders.[43] In the private sector, insurance companies use personal data with embedded biases to set rates and determine risk, unjustly rewarding some groups and penalizing others, and marketers exploit the behavioral habits of people with data that reinforces those habits and appeals to personal impulses.[44]

Law enforcement often relies on private data companies to obtain information that it cannot legally obtain otherwise. Profiling the life and death of Hank Asher, a former drug smuggler who became a pioneer in data broker services, *ProPublica* and *The New York Times* reporter McKenzie Funk acknowledged the ways in which the algorithmic processing of personal data can go wrong, encoding social biases into its code, making predictions about events that occur so rarely that it often produces false positives, and concealing its functions in a black box that is difficult for outsiders to inspect. However, Funk expressed more concern about when these algorithms do work, reinforcing models of living based upon past actions instead of future potential. Funk writes about the use of personal data: "And insofar as it cripples social mobility because you're stuck in your own pattern, it could further hasten the end of the American dream."[45]

The amount of personal data that can be used for these purposes is steadily increasing. The management consulting firm McKinsey forecasts that the use of digital data to predict the wants and needs of consumers will drive demand for "systems that can pool and analyze structured and unstructured data, algorithms that can identify behavioral patterns and customer propensity, and analysis capabilities to feed that information into easy-to-use dashboards."[46] The data collected includes the digitization of physical space, including using GPS data to locate customers and predict their movements. Another set of data points includes measuring and predicting the emotional states of individuals. An entire field called affective computing has emerged in which the seemingly irreducible range of human emotion can be registered and stored digitally. Computers can simulate emotions in their text and voice output, and AI algorithms have been trained to recognize the emotional states of humans through their voice. Even social relationships can be quantified and stored: the analysis of uploaded family photographs can reveal the strength of social ties and kinship that influence affective states.[47]

The inclusion of VR data into the personal information matrix can provide much more direct measurements of affective states and generate more accurate simulations of emotions expressed by digital avatars interacting with users. While people can suppress their internal emotional state and present a constructed identity to the outside world for social consumption, biometric data, including pulse, eye movements, pupil dilation, and pace and patterns of movement, are more difficult to conceal. Like mobile computing, VR-related spatial computing technology can provide real-time information about the location of individuals and where they have directed their attention. Having access to this data and acting on it provides large companies with a level of power that many people are unaware is possessed by them; in fact, their own performative stances can be incorporated into the personal data stream for further insights into what emotions and actions users are attempting to conceal. The imbalance of power between users and those able to collect and process the massive amount of data about them presents ethical challenges that test conceptions of autonomy, personal agency, and the equitable distribution of technological benefits from the algorithmic manipulation of data.

A human right to privacy can serve as a bulwark against these imbalances, but this right is not absolute and varies widely in different countries. Recently, 12 states in the United States have adopted comprehensive privacy laws that protect personal or sensitive information from being disseminated by businesses without the consent of consumers by opting in for that collection or clearly offering a chance to opt out.[48] However, many other states do not have a comprehensive policy, with some using the lack of regulation to attract businesses seeking to avoid barriers restricting their use of personal data.

In June 2023, the European Union released draft legislation titled the AI Act to react to the quickly developing field of AI. Much of the proposed law, though, focuses specifically on the collection and analysis of personal data. For example, the legislation identifies as an unacceptable risk the use of social scoring, which it clarifies as "classifying people based on behaviour, socio-economic status or personal characteristics."[49] The law would also prohibit "cognitive behavioural manipulation of people or specific vulnerable groups" and "real-time and remote biometric identification systems."[50] The EU also seeks to address the transparency of AI use by properly labeling content generated by these algorithms. To date, this is the earliest and most stringent attempt to regulate AI, but the law would also set standards for governing the use of personal data in VR and other digital applications.

Other governments have used the philosophy of state security instead of individual rights when examining the limits of privacy. India has put in place the Digital Personal Data Protection Act, ensuring that citizens have the right to examine, correct, or delete the personal data companies collect about them. The law also provides special protections regarding the data of children and other vulnerable groups. However, critics claim the law also gives the government sweeping new powers to use personal data to surveil its citizens and to censor digital content that does not support its policies. In particular, critics say the law threatens the free press in India, allowing government officials to investigate journalists and prevent them from accessing and publishing information about governmental activities.[51]

Over the past decade, China has been implementing what it calls a social credit score, an evaluation of each citizen based on their financial state, legal history, past Internet searches, and political activity, among other data sources. A person's score may help determine not only the ability to secure a loan or government position but may also be used to restrict travel and access to the court system.[52] A report for the U.S.-China Economic and Security Review Commission found that this social scoring lacks transparency and is assisting the central Chinese Communist Party in exercising greater control over local governments. New technologies like facial recognition and remote sensors help to enforce restrictions put in place by the social credit score, and both domestic and international companies are monitored without a means to inspect or challenge decisions made as a result of their social credit rating. Through the fear of a bad score, the central government can control the behavior of its citizens and effectively blacklist citizens and companies that deviate from the central government's policies and messaging.[53] The sublimation of individual rights of privacy to state security helps to build a totalitarian society by using data collection and analysis to enforce with greater precision existing and future laws.

While the United States considers new laws and regulations to protect the privacy of citizens from the dissemination of their personal data, some agencies continue to collect and analyze the information of others for purposes of national security. The National Security Agency's PRISM program is set up for mass collection of the communications of foreigners even if that data is on U.S. soil and contains information about U.S. citizens. Whistleblower NSA contractor Edward Snowden exposed the secret program in 2013, but the program continues to operate under the legal protection of Section 702 of the Foreign Intelligence Surveillance Act. While the NSA insists that no American citizens are targeted in the program, the FBI searched the database over 3.4 million times in 2021 alone, according to a report by the Director of National Intelligence.[54]

Given the ability of governments and large companies to collect, store, process, and use personal data for the purposes of predicting and controlling behavior, the ability of VR to provide even more personal data that reveals real-time biometric and affective data, and the tensions between the right to privacy, the right to information, and the right to security, how should an ethics of VR manage the balance between information sharing and data security? If, as Internet pioneer Stewart Brand said, "information wants to be free," does complete transparency solve the issue of widespread dissemination of personal information, or should there be some data that remains off-limits to everyone other than the person it references?

One group that current privacy laws commonly address is children. Recognized by several governments as a protected class, the personal data of children is often cited as a reason to hold companies back from widespread collection. Epic Games, the maker of the popular video game Fortnite, recently agreed to pay $275 million in fines and refund about $245 million to families for violating the Children's Online Privacy Protection Act. Federal Trade Commission Lina M. Khan remarked about the settlement, "Protecting the public, and especially children, from online privacy invasions and dark patterns is a top priority for the Commission."[55] While prosecutors protect children by targeting traditional video games played on a screen that gather data, schools are readily adopting VR as a necessary skill to teach children for a future immersive workforce. Girard Kelly, the Director of Privacy for the nonprofit group Common Sense, concluded that the organization cannot "recommend any device right now for schools and districts that wouldn't potentially be violating any state or federal privacy laws."[56] Their analysis showed that VR headsets display third-party advertising to children while gathering biometric data that is transmitted back to servers, where it may be collected and sold. Immersive technology consultant Emory Craig argues that mixed reality, combining virtual elements with a view of the physical world, poses an even greater threat to privacy, as the same data gathered in VR can be applied to the real-world elements with which a user is interacting. This gives marketers and government enforcement agencies alike unprecedented insights into an individual's behavior, interactions, and habits.[57]

Among other protections, the United Nations Universal Declaration of Human Rights addresses the universal right to privacy. Article Twelve states, "No one shall be subjected to arbitrary interference with his privacy, family, home or correspondence, nor to attacks upon his honour and reputation. Everyone has the right to the protection of the law against such interference or attacks."[58] Elsewhere, the declaration asserts the right to the freedom of thought and movement. If these rights are to be guaranteed, they must be protected from erosion by private and public interests. Writing about Facebook as a problem for democracy, information technology scholar Siva Vaidhyanathan argues in his book *Antisocial Media*, "Protecting privacy is an environmental problem that Facebook treats as a matter of individual responsibility."[59] As the maker of the most widely distributed VR headset, Meta emphasizes individual freedom and responsibility for managing user privacy preferences. However, without a clear and transparent way of understanding how data is being gathered, used, and sold, only a systematic ethics of VR privacy can address the power imbalance between the data brokers and the VR users that the data represents.

NOTES

1. Rosenberg, Matthew, Nicholas Confessore, and Carole Cadwalladr. "How Trump Consultants Exploited the Facebook Data of Millions." *The New York Times*, March 17, 2018. https://www.nytimes.com/2018/03/17/us/politics/cambridge-analytica-trump-campaign.html.

2. Cadwalladr, Carole, and Emma Graham-Harrison. "Revealed: 50 Million Facebook Profiles Harvested for Cambridge Analytica in Major Data Breach." *The Guardian*, March 17, 2018. https://www.theguardian.com/news/2018/mar/17/cambridge-analytica-facebook-influence-us-election.

3. Kaiser, Brittany. *Targeted: My Inside Story of Cambridge Analytica and How Trump, Brexit and Facebook Broke Democracy*. London: HarperCollins, 2019, 327.

4. Grewal, Paul. "Suspending Cambridge Analytica and SCL Group from Facebook." *Meta* (blog), March 17, 2018. https://about.fb.com/news/2018/03/suspending-cambridge-analytica/.

5. Bowman, Emma. "After Data Breach Exposes 530 Million, Facebook Says It Will Not Notify Users." *NPR*, April 9, 2021. https://www.npr.org/2021/04/09/986005820/after-data-breach-exposes-530-million-facebook-says-it-will-not-notify-users.

6. Federal Trade Commission. "FTC Imposes $5 Billion Penalty and Sweeping New Privacy Restrictions on Facebook." Federal Trade Commission, July 24, 2019. https://www.ftc.gov/news-events/news/press-releases/2019/07/ftc-imposes-5-billion-penalty-sweeping-new-privacy-restrictions-facebook.

7. Hu, Margaret. "Cambridge Analytica's Black Box." *Big Data & Society* 7, no. 2 (July 2020): 205395172093809. https://doi.org/10.1177/2053951720938091.

8. Altman, Sam. "Planning for AGI and Beyond." *OpenAI*, February 24, 2023. https://openai.com/blog/planning-for-agi-and-beyond.

9. OpenAI. "The Power of Continuous Learning." *OpenAI*, December 23, 2022. https://openai.com/blog/the-power-of-continuous-learning.

10. Smith, Craig S. "Hallucinations Could Blunt ChatGPT's Success." *IEEE Spectrum*, March 13, 2023. https://spectrum.ieee.org/ai-hallucination.

11. Thorbecke, Catherine. "OpenAI Launches a Version of ChatGPT for Businesses." *CNN*, August 28, 2023. https://www.cnn.com/2023/08/28/tech/chatgpt-enterprise-openai/index.html.

12. Ludlow, Edward. "OpenAI Nears $1 Billion of Annual Sales as ChatGPT Takes Off." *Bloomberg.Com*, August 29, 2023. https://www.bloomberg.com/news/articles/2023-08-30/openai-nears-1-billion-of-annual-sales-as-chatgpt-takes-off.

13. Gurman, Mark, and Julia Love. "Apple Nears Deal With OpenAI to Put ChatGPT on iPhone." *Bloomberg.com*, May 10, 2024. https://www.bloomberg.com/news/articles/2024-05-11/apple-closes-in-on-deal-with-openai-to-put-chatgpt-on-iphone.

14. Patel, Nilay. "Google and YouTube Are Trying to Have It Both Ways with AI and Copyright." *The Verge*, August 22, 2023. https://www.theverge.com/2023/8/22/23841822/google-youtube-ai-copyright-umg-scraping-universal.

15. Mauran, Cecily. "Apple Vision Pro: ChatGPT Is Now Available on the $3,500 Headset." *Mashable*, February 6, 2024. https://mashable.com/article/apple-vision-pro-chatgpt-visionos-app.

16. Jeff Orlowski, dir. *The Social Dilemma*. Exposure Labs, 2020. https://netflix.com/title/81254224.

17. Leach, William. *Land of Desire: Merchants, Power, and the Rise of a New American Culture*. New York: Vintage Books, 1994, 37.

18. Newport, Frank. "75 Years Ago, the First Gallup Poll." *Gallup.com*, October 20, 2010. https://news.gallup.com/opinion/polling-matters/169682/years-ago-first-gallup-poll.aspx.

19. The Business Research Company. "Market Research Services Global Market Briefing 2023." *Marketresearch.com*, January 2023. https://www.marketresearch.com/Business-Research Company-v4006/Research-Services-Global-Briefing-33140905/.

20. Fowler, Geoffrey A. "The Spy in Your Wallet: Credit Cards Have a Privacy Problem." *Washington Post*, August 29, 2019. https://www.washingtonpost.com/technology/2019/08/26/spy-your-wallet-credit-cards-have-privacy-problem/.

21. For information about this service, see: "Customer Intelligence Cloud." *Acxiom*. https://www.acxiom.com/customer-intelligence/cloud/.

22. Tzortzis, Demetrios. "Diner Recognition Help Available for the QSR Industry." *Acxiom*, August 2, 2023. https://www.acxiom.com/blog/from-data-to-dollars-how-diner-recognition-can-boost-restaurant-and-qsr-sales/.

23. Epsilon Data Management, LLC. "Understand Your Customer Analytics with Consumer Data." Accessed September 10, 2023. https://www.jpgilon.com/us/products-and-services/data.

24. Microsoft. "Everything You Need to Know About Internet Cookies." *Edge*. Accessed September 10, 2023. https://www.microsoft.com/en-us/edge/learning-center/what-are-cookies.

25. The International Association of Privacy Professionals. "Web Beacon." Accessed September 10, 2023. https://iapp.org/resources/article/web-beacon/.

26. Weinshel, Ben, Miranda Wei, Mainack Mondal, Euirim Choi, Shawn Shan, Claire Dolin, Michelle L. Mazurek, and Blase Ur. "Oh, the Places You've Been! User Reactions to Longitudinal Transparency About Third-Party Web Tracking and Inferencing." In *Proceedings of the 2019 ACM SIGSAC Conference on Computer and Communications Security*, 149–166. London United Kingdom: ACM, 2019. https://doi.org/10.1145/3319535.3363200.

27. Acxiom. "What Is Identity Resolution?" Accessed September 10, 2023. https://www.acxiom.com/identity-resolution/.

28. Burgess, Matt. "The Quiet Way Advertisers Are Tracking Your Browsing." *Wired*, February 26, 2022. https://www.wired.com/story/browser-fingerprinting-tracking-explained/.

29. Atske, Sara. "Americans and Privacy: Concerned, Confused and Feeling Lack of Control Over Their Personal Information." *Pew Research Center: Internet, Science & Tech* (blog), November 15, 2019. https://www.pewresearch.org/internet/2019/11/15/americans-and-privacy-concerned-confused-and-feeling-lack-of-control-over-their-personal-information/.

30. Lomas, Natasha, and Romain Dillet. "Terms and Conditions Are the Biggest Lie of Our Industry." *TechCrunch* (blog), August 21, 2015. https://techcrunch.com/2015/08/21/agree-to-disagree/.

31. Meta Platforms, Inc. "Meta Privacy Policy—How Meta Collects and Uses User Data." *Meta Privacy Center*, December 27, 2023. https://www.facebook.com/privacy/policy/.

32. Miller, Mark Roman, Fernanda Herrera, Hanseul Jun, James A. Landay, and Jeremy N. Bailenson. "Personal Identifiability of User Tracking Data during Observation of 360-Degree VR Video." *Scientific Reports* 10, no. 1 (October 15, 2020): 17404. https://doi.org/10.1038/s41598-020-74486-y.

33. Zuboff, Shoshana. *The Age of Surveillance Capitalism: The Fight for a Human Future at the New Frontier of Power*. New York: PublicAffairs, 2020.

34. Zimmerli, Lukas, Mario Jacky, Lars Lünenburger, Robert Riener, and Marc Bolliger. "Increasing Patient Engagement During Virtual Reality-Based Motor Rehabilitation." *Archives of Physical Medicine and Rehabilitation* 94, no. 9 (September 2013): 1737–1746. https://doi.org/10.1016/j.apmr.2013.01.029.

35. Felsberg, Danielle T., Jaclyn P. Maher, and Christopher K. Rhea. "The State of Behavior Change Techniques in Virtual Reality Rehabilitation of Neurologic Populations." *Frontiers in Psychology* 10 (May 8, 2019): 979. https://doi.org/10.3389/fpsyg.2019.00979.

36. Freeman, Daniel, Polly Haselton, Jason Freeman, Bernhard Spanlang, Sameer Kishore, Emily Albery, Megan Denne, Poppy Brown, Mel Slater, and Alecia

Nickless. "Automated Psychological Therapy Using Immersive Virtual Reality for Treatment of Fear of Heights: A Single-Blind, Parallel-Group, Randomized Controlled Trial." *The Lancet Psychiatry* 5, no. 8 (August 2018): 625–632. https://doi.org/10.1016/S2215-0366(18)30226-8.

37. King, Barbara J. "Can Cute Virtual Reality Spiders Help Reduce Arachnophobia?" *NPR*, September 1, 2016. https://www.npr.org/sections/13.7/2016/09/01/491991386/can-cute-virtual-reality-spiders-help-reduce-arachnophobia.

38. Miloff, Alexander, Philip Lindner, Peter Dafgård, Stefan Deak, Maria Garke, William Hamilton, Julia Heinsoo, et al. "Automated Virtual Reality Exposure Therapy for Spider Phobia vs. in-Vivo One-Session Treatment: A Randomized Non-Inferiority Trial." *Behaviour Research and Therapy* 118 (July 2019): 130–140. https://doi.org/10.1016/j.brat.2019.04.004.

39. Bellinger, Daniel, Kristin Wehrmann, Anna Rohde, Maria Schuppert, Stefan Störk, Michael Flohr-Jost, Dominik Gall, et al. "The Application of Virtual Reality Exposure versus Relaxation Training in Music Performance Anxiety: A Randomized Controlled Study." *BMC Psychiatry* 23, no. 1 (August 1, 2023): 555. https://doi.org/10.1186/s12888-023-05040-z.

40. *Carpenter v. United States*, 585 U.S. (2018).

41. Goitein, Elizabeth. "The Government Can't Seize Your Digital Data. Except by Buying It." *Washington Post*, April 27, 2021. https://www.washingtonpost.com/outlook/2021/04/26/constitution-digital-privacy-loopholes-purchases/.

42. O'Neil, Cathy. *Weapons of Math Destruction: How Big Data Increases Inequality and Threatens Democracy*. New York: Crown, 2016.

43. Angwin, Julia, Jeff Larson, Surya Mattu, and Lauren Kirchner. "Machine Bias." *ProPublica*, May 23, 2016. https://www.propublica.org/article/machine-bias-risk-assessments-in-criminal-sentencing.

44. Schneier, Bruce. *Data and Goliath: The Hidden Battles to Collect Your Data and Control Your World*. New York: W.W. Norton & Company, 2015.

45. Funk, McKenzie. "The Man Who Trapped Us in Databases." *The New York Times*, September 22, 2023. https://www.nytimes.com/2023/09/22/magazine/hank-asher-data.html.

46. Boudet, Julien, Brian Gregg, Kathryn Rathje, Eli Stein, and Kai Vollhardt. "The Future of Personalization—and How to Get Ready for It." *McKinsey*, June 18, 2019. https://www.mckinsey.com/capabilities/growth-marketing-and-sales/our-insights/the-future-of-personalization-and-how-to-get-ready-for-it.

47. *ASMMC-MMAC'18: Proceedings of the Joint Workshop of the 4th Workshop on Affective Social Multimedia Computing and First Multi-Modal Affective Computing of Large-Scale Multimedia Data*. New York: Association for Computing Machinery, 2018.

48. Schapiro, Zachary S. "Update: Processing Sensitive Personal Information under U.S. State Privacy Laws." *The National Law Review*, September 12, 2023. https://www.natlawreview.com/article/update-processing-sensitive-personal-information-under-us-state-privacy-laws.

49. European Parliament. "EU AI Act: First Regulation on Artificial Intelligence," June 8, 2023. https://www.europarl.europa.eu/news/en/headlines/society/20230601STO93804/eu-ai-act-first-regulation-on-artificial-intelligence.

50. European Parliament. "EU AI Act: First Regulation on Artificial Intelligence."

51. Castro, Chiara. "India Approves New Privacy Law, but Is It Really a Win for Citizens?" *TechRadar*, August 24, 2023. https://www.techradar.com/computing/cyber-security/india-approves-new-privacy-law-but-is-it-really-a-win-for-citizens.

52. Chin, Josh, and Gillian Wong. "China's New Tool for Social Control: A Credit Rating for Everything." *Wall Street Journal*, November 28, 2016. http://www.wsj.com/articles/chinas-new-tool-for-social-control-a-credit-rating-for-everything-1480351590.

53. Schaefer, Kendra. "China's Corporate Social Credit System." *Trivium China*, November 16, 2020. https://www.uscc.gov/sites/default/files/2020-12/Chinas_Corporate_Social_Credit_System.pdf.

54. Kaplan, Fred. "Why Biden Wants to Keep the Law That Allows NSA Mass Surveillance, and Republicans Want to Kill It." *Slate*, March 1, 2023. https://slate.com/news-and-politics/2023/03/why-biden-wants-to-keep-section-702-nsa-mass-surveillance.html.

55. Federal Trade Commission. "Fortnite Video Game Maker Epic Games to Pay More Than Half a Billion Dollars over FTC Allegations of Privacy Violations and Unwanted Charges." *Federal Trade Commission*, December 19, 2022. https://www.ftc.gov/news-events/news/press-releases/2022/12/fortnite-video-game-maker-epic-games-pay-more-half-billion-dollars-over-ftc-allegations.

56. Klein, Alyson. "VR Devices Collect 'Intimate' Data, Lack Privacy Protections. Should Schools Invest?" *Education Week*, November 18, 2022. https://www.edweek.org/technology/vr-devices-collect-intimate-data-lack-privacy-protections-should-schools-invest/2022/11.

57. Craig, Emory. "Think Oculus Rift and Privacy Is an Issue? Just Wait until Mixed Reality Arrives," April 11, 2018. https://www.digitalbodies.net/oculus-rift-and-privacy-issue-just-wait-until-mixed-reality-arrives/.

58. United Nations. "Universal Declaration of Human Rights." Accessed September 17, 2023. https://www.un.org/en/about-us/universal-declaration-of-human-rights.

59. Vaidhyanathan, Siva. *Antisocial Media: How Facebook Disconnects US and Undermines Democracy*. New York: Oxford University Press, 2018, 74.

Chapter 6

Misrepresentation and Bias in VR

Virtual reality (VR) games and fiction narratives allow users to escape to other worlds, real or imagined. In addition, immersive documentaries, news, and live entertainment are expanding genres of 360-degree video, all viewed within a VR headset. In the meantime, gaming has become largely a social activity, with live chats, multiplayer games, and E-sports leagues all becoming popular over the last two decades. Meta Platforms has made a substantial investment in promoting social VR, in which users interact with others within immersive environments, and Apple promotes using its Vision Pro device to communicate with other users who are represented by their avatars. All of these uses of VR have the potential for new representations and discovery, but they also can carry with them and perpetuate existing social biases and misrepresent the intentions and experiences of people engaging with others within immersive environments.

When a person is represented by an avatar in VR, that avatar can shape how other users perceive that person. An ethics of VR can provide guidelines as to when users are obligated to be transparent in representing who they are and under what conditions they can be allowed the freedom to alter their appearance and other personal information. Different virtual environments each have their own implicit rules about concealing or revealing one's identity or trying on alternative identities. These issues are informed by earlier debates over virtual communities as they formed with the growth of the Internet, as well as more recent concerns over deepfakes, catfishing, AI-generated content, and other products of malicious actors.

Even when developers and users have sincere intentions, implicit biases may distort the representation of people and groups in VR. Though the limits of geography are removed and a degree of anonymity can be achieved, virtual communities often are projections of preexisting communities in the physical world. People are still behind keyboards or VR headsets when they engage in virtual communities, and they bring with them ingrained biases developed over past experiences and social conditioning.

Designers as well have the ability to manipulate the representation of virtual environments. In fact, a benefit of VR is the ability to present environments that deviate from everyday experiences. Users can explore outer space, alien worlds, underwater cities, or fly through the sky, all from the safety of their living rooms. However, this capability can be abused. VR documentaries, aided by AI, may falsify events for political or commercial gain while creating a feeling of real events through their sensory immersion. Ultimately, the confusion of real and manipulated worlds erodes the trust users place in the medium, or any medium for that matter. An ethics of VR can spell out the obligation of VR designers to represent the physical world accurately or make clear when virtual worlds deviate from the real world.

An earlier experimental technology may offer insights into the issues of representation for VR: Google Glass. Google attempted to market a pair of eyeglasses with a thin frame, small prismatic lenses, and large arms that contained the battery, processing unit, and wireless transceiver. These glasses, in fact, were an early form of AR technology. People could take digital photos while wearing them, much like Meta's Ray-Ban Stories sunglasses can do today.[1] In addition, wearers of Google Glass could browse the Internet and search for information using voice commands, with the results displayed translucently over the view of the physical world. Because the glasses included a camera that could surreptitiously record anyone or anything within view, including unsuspecting strangers, sensitive documents, and live performances, Google Glass prompted concerns about whether the technology would be used to steal copyrighted information, circumvent security, or encourage the adoption of a surveillance society.[2] Researchers Safiya Umoja Noble and Sarah T. Roberts expose the contradiction underlying the fact that it was the wealthy and privileged who could afford to have access to this technology, allowing this group to help impose a surveillance state through their ability to watch and record people and events around them. This localized and personal form of surveillance benefits those conducting the surveillance but disproportionately hurts those with less privilege. They write:

> Yet undeniably, Google Glass has been deployed as a project in service of power elites (those less affected by hyper-surveillance), yet the consequences of normalizing the loss of privacy via heightened surveillance, the differing consequences of surveillance on people dependent on race, socioeconomic status, and other identity-related factors, and the inability for all to equally access technologies like Glass lead to a disregard for the types of harm that are disproportionately impacting poor people, people of color, and women.[3]

Ultimately, Google Glass was not successful for a myriad of reasons, including the fact that the public simply may not have been ready for such wearable technology, especially when only a few could afford to wear it. Ray-Ban Stories, however, continues to be sold under the more friendly banner of a social media tool, allowing its wearers to share what they record with their networks as a form of entertainment. Apple's Vision Pro, marketed as a productivity tool, or "spatial computing," according to Apple's website, also has the capability to record its environment, but it goes a step further by digitizing it and presenting that view to users within its three-dimensional display.[4] While this technology is used for presenting the outside world as it exists with digital information overlaid on this image, this method of display makes it possible to modify the appearance of the real environment so that users see the modification and not the image that the device's cameras see.

The evolving capability of devices such as these raises important questions about using these technologies to interact with physical reality, including questions about who most likely has access to these technologies, how VR technologies may modify the view of the environment or other people, and what people using these technologies will do with the knowledge gained from augmented views of other people and environments. Even more so, if control is ceded to others regarding the manner of displaying people and environments in VR, how could VR misrepresent these representations, and what biases will become embedded in these representations that may be mistaken for reality? These questions necessarily require an ethical analysis of the intersections between race, gender, class, and identity with VR, AR, and XR, particularly focusing on how these technologies might affect those most vulnerable to their influences.

BEYOND THE BODY

For many people, VR presents the first opportunity to transcend their own body and inhabit the body of someone (or something) else. In fact, VR experiences are often advertised in this way. At the beginning of a new game, players often are presented with a menu of faces, bodies, skins, facial hair, and clothing to choose from, similar to what screened video games on the market have been doing for some time. When users set up an account in the Meta Quest, they can choose an avatar that includes several different skin tones, hair, and genders. (See figure 6.1.)

In some ways, this certainly is an improvement over the past, when whiteness was a default for photographic representation. Lorna Roth explores the historical racial bias of photography by examining the color balance of the film medium itself in "Looking at Shirley, the Ultimate Norm: Colour

Figure 6.1. Setting Up an Avatar in Meta Quest. *Screenshot by author.*

Balance, Image Technologies, and Cognitive Equity." She documents how the photographic film camera and photo development design used white models, all given the name "Shirley," as the industry standard. In turn, this process made white faces the dominant group that photography was positioned to capture and designed to highlight. People with darker skin, in turn, were presented in photographs with darker tones and less detail than they appeared in reality. The photograph, in essence, favored white subjects over Black ones. Roth argues, "I believe that the potential building blocks of cognitive equity will be located in small and subtle changes in our taken-for-granted perceptions and behaviours, resulting from and active demand for a wider range of socially imagined possibilities for inclusiveness."[5] Long before the development of VR, the history of photography demonstrates that, from the very beginning of mechanical image reproduction, systematic bias became an inherent part of the medium. This historical lesson should caution developers of new imaging technologies to attend to and correct for latent biases that can favor one group over another in the medium. This holds especially true for VR, which, in its name itself, makes a promise that it presents a digital symbolic reality that approaches the immersion of physical reality more so than any other medium existing before it.

For some people, the process of picking out an avatar to represent them is a momentary decision. For others, however, it requires careful thought and decision-making. Numerous sites, blogs, and Reddit discussions provide advice on the art of choosing an identity for the virtual world. Researchers have discovered that people use a variety of strategic approaches to avatar selection. Kristine L. Nowak and Jesse Fox write,

> Some users will select avatars that accurately represent something about them, but that inaccurately present other aspects of the self. While this may sometimes be a choice, other times it is not. Some digital environments make it difficult to convey one's identity authentically due to technological constraints or social norms.[6]

Some people recognize the opportunity they have when they find themselves in a powerful position to decide what type of body they want. However, this power comes with both limits and responsibility.

While avatars can provide people with the realization of their desire to construct their own identity (at least in virtual environments), too often the choice consists of limited options that merely mimic currently available representations that already exist in public life. In addition, while looking at something as basic as the skin color of a character, interesting debates necessarily arise about what the limits of choices a person may have and the question of who should have the right or not to select a skin color not their own. For example, is it appropriate for a white user in the United States immersed in stereotypes about other cultures, to represent themselves as a Black character, or for a white European user to take on the appearance of an Indian or East Asian character? What are the limits in the performative aspects of adopting an avatar that should not be transcended, and who should be able to decide what those limits are?

Surely, the intent of the user must have some influence over the answer to these questions. Someone from a dominant class and ethnicity who takes on the identity of an oppressed group in order to play out the stereotypes of that group is quite different from a user who takes on another identity to witness how others in virtual environments treat them. Likewise, a member of an oppressed group who takes on the identity of a dominant group may discover how they are treated differently. The type of role-playing users engage in can help to build empathy or close it off. Researchers found that white players taking on Black avatars in violent video games often displayed more aggressive attitudes and displayed negative stereotypes about Black players, which in turn reinforced their implicit biases against Black players.[7] Creators of virtual worlds have a responsibility to promote empathy over reinforcing stereotypes, but the question of how much constraint they should exercise over allowing users to adopt different appearances remains open.

The player in a video game may think their choice over which character they use to play is ultimately unimportant. Whether someone chooses a character that is completely imaginary, like a blue dinosaur or a purple alien with three heads, or they build an avatar that represents some aspect of who they are in real life, who they wish to be, or what face they would like to put on as a costume or disguise, they are able to shut the game off at the end of their time playing it. As the physical and virtual worlds continue to overlap, however, the ability to simply shut off a digital representation of a user will not be as easy a task. Scholars of digital environments predict that the physical and virtual worlds will not remain distinct entities but will increasingly merge so that people will inhabit both simultaneously without making delineations between the two. Marketers are working on a future when people can share their customized realities with others and be presented with products and services that help support that constructed reality.[8]

A person's existence in this constructed reality likely will not be constructed entirely from the desires of each individual; elements borrowed from the physical world will be replicated and limit what is possible. While greater flexibility is possible in virtual environments, they likely will be grounded to some degree in the physical world. Digital Twinning is a technology being developed that creates avatars of users which match each corresponding user in the real world as they enter the virtual environment. Technology and marketing consultants Luis Bravo Martins and Samantha G. Wolfe write: "As the end goal of the metaverse is to synchronize digital and physical, then the prevalent digital format will most likely be 3D."[9] Technologies like three-dimensional scanning, NeRF, and motion capture already allow for the capture and virtual reproduction of photorealistic faces and bodies, and real-time rendering of these figures will soon be achievable. If the value of reproducing the real world in the metaverse wins out over the desire for opening up new possibilities of representation, it is quite possible that VR will replicate existing inequalities and biases instead of helping to imagine a post-racialized and classless society.

VIRTUAL BLACK FACE

The singer Adele created a public relations problem after her visit to the Notting Hill Carnival in 2020. While attending, she wore a Jamaican flag bikini and had her hair in Bantu knots.[10] What she initially thought of as supporting a historically oppressed culture was widely received and criticized as appropriation of that culture as a fashion statement. While this controversy occurred in the real world, what are the boundaries of appropriation when the world you exist in is not real or the players you become are not you?

In addressing the issue of cultural appropriation, University of Connecticut Professor of Law Willajeanne F. McLean asks: "The sixty-four-thousand dollar question is who has the right to represent another culture? Who has the authority to use that culture or its features, defined broadly, for commercial gain?"[11] The Adele controversy and the social media debate that it sparked make it clear that many people in online communities are becoming more knowledgeable about and wrestling with issues of cultural appropriation. An ethics of VR needs to address these questions because of its ability to allow for immersive interactions using constructed digital representations and performances that can deviate from physical reality but most likely reproduce it in one form or another.

The company Mursion seemed to be surprised to find itself in the middle of a controversy over its diversity training videos. Originating as a company created to provide teacher training tools, Mursion had evolved to offer corporate clients VR simulation tools focused on educating employees about race and diversity issues. However, when employees logged onto their training sessions, they heard white actors voicing Black characters in some of the simulations. Rather than hire Black actors to realistically play Black employees in the animated videos, the white actors used somewhat exaggerated Black dialects in what some listeners described as a caricature. The company was engaging in an audio version of Blackface, a widely condemned practice of having white actors represent Black characters.[12]

In a chapter of the book *Ethics of Emerging Media*, Elspeth Tilley discusses the ethical concept of consequentialism. She describes its goal as one in which the impacts of a decision are evaluated within an interdependent environment. She writes, "To effectively practice outcome-based ethics, one must seek to develop simultaneous awareness of all of the audiences of one's actions, all of the interconnected impacts of one's behavior."[13] Though Tilley focuses on ethics in the area of public relations, this idea also can be applied to the ethics of behavior in virtual worlds, especially when considering bias and the representation of race. People might be focused on their own actions and perspective when they choose an avatar and do not understand that virtual identity as a real representation of themselves. However, the selves they inhabit help to construct real social groups. These choices and their consequences should be considered when adopting a particular racialized character, and their behavior while embodying that role is a component of building an environment that either deconstructs or reinforces stereotypes and biases among everyone in that social group.

While an ethics of VR can safely categorize the appropriation of another culture for profit that results in the ridicule of that culture as an undesirable condition, what becomes less clear is the ethics of VR apps allowing users to take on the appearance of a member of a different cultural group for the purpose of learning or empathy. For example, Body of Mine allows VR users to imagine what it would be like if they inhabited the body of a different gender. The immersive experience instructs users to touch parts of their body and hear recordings of trans people talking about their own experiences and relationships with their bodies. The purpose, according to the creators of the experience, is to educate cis-gendered people on what is it like to experience gender dysphoria.[14] While the purpose of the project is to generate awareness and empathy, its success is partially determined by the context in which the app is experienced and the intent of the user. How can an ethics of VR address these different contexts to ensure outcomes that help reduce bias and structural social inequalities?

VIRTUAL WORLDS AS SITES OF ADVOCACY

The ethics of care is an approach that originates from feminist theory. Carol Gilligan and Nel Noddings were leaders in developing this shift from traditional deontological ethics and consequentialism. This form of ethics is based on the fact that emotions, which often are discouraged from being considered in ethics, can be useful in decision-making. They argue that relationships, social bonds, and people without embedded power need to be taken into consideration when making ethical decisions.[15] Instead of attempting to approach ethics from an objective or neutral point of view, who is being cared for is central to the decision-making process under an ethics of care framework. Carolina Villegas-Galaviz and Kirsten Martin explore the value of using ethics of care as a model in considering power and decision-making processes for AI technology, arguing that using the theories developed by Gilligan and Noddings highlights empathy and context in ethical decision-making. They write,

> Within the study of the ethics of care, there exists a shared understanding that ethical decision-making should consider *interdependent relationships, context and circumstances,* attend to particular *vulnerabilities,* and also should respond to different points of view, and hear different *voices.*[16]

People tend to assume that technology derived from computers and machines is neutral and objective. By focusing on the people that experience the changes these technologies enact, particularly those most vulnerable to these changes, an ethics of care offers possibilities for centering empathy and a concern for maintaining the interdependence among people at the forefront of technology use.

One possible use of theorizing an ethics of care in VR is considering the complicated relationship between those who are making the experiences and those who are experiencing them. The nascent field of immersive journalism is one area where these relationships are foregrounded. VR pioneers like Nonny de la Peña have long advocated for VR experiences that promote empathy. Engaging in immersive journalism, her VR film *One Dark Night* uses court records and the recording of a 911 call to tell the story of the killing of Trayvon Martin, a Black teenager, by neighborhood watch volunteer George Zimmerman in Florida.[17] The VR animation, paired with real audio recordings, allows for viewing from multiple perspectives in the neighborhood while combining a variety of media. The experience extends the senses of users on multiple levels and increases the sense of immersion. De la Peña argues that this immersion is the key to opening people up to diverse perspectives in order to become more informed citizens in a participatory democracy.

Some developers think that VR can dismantle racist structures by allowing people from dominant groups to experience racism themselves. For example, MIT researchers created a VR experience called "On the Plane," which allows people to take on the perspective of different characters on an airplane flight. One character is Muslim-American and experiences discrimination, another is a white passenger engaged in the discrimination, and a third character is a bystander witnessing the discrimination that takes place.[18] Likewise, "1000 Cut Journey," which premiered at the 2018 Tribeca Film Festival, offers a visceral experience of racism by having users embody a character who experiences incidents from childhood through adulthood at home, in the classroom, from the police, and at the workplace.[19]

Other creators are experimenting with using VR as a way to educate potential jurors and judges about implicit bias. This form of bias occurs when individuals internalize unconscious preconceptions about people of color that, in turn, impact how they treat and judge members of oppressed groups. In a VR experience created at Stanford University, students were given an implicit bias test before being asked to wear a VR headset that allowed them to experience the body ownership illusion. This illusion involves watching in the HMD an avatar's body being touched while the user's physical body is also touched simultaneously. The addition of tactile stimulation, reinforcing the visual illusion in VR, increases the impact of the immersive experience.[20] While researchers hoped to universally reduce implicit bias with the use of VR, the results were more complicated. The reduction varied depending on the physical body of the user, the type of avatar that was embodied, and the initial knowledge of those engaged in the exercise.[21] However, the authors concluded that: "As a preliminary investigation into VR and implicit bias in the courtroom . . . embodying an outgroup avatar for five minutes can have a positive impact on implicit racial biases."[22]

VR experiences do not have to be limited to social situations or conflicts that represent the present day. Because immersive experiences can simulate not just any space but also spaces in different times, they can enable people to experience simulations of historical discrimination. The VR experience “Barnstormers: Determined to Win” allows players to experience discrimination by embodying men of color during the early days of segregated baseball teams. “Traveling While Black” is a VR experience that introduces users to the restrictions and discrimination Black travelers faced in the South during the Jim Crow era. In “Mapping Amache,” users can virtually tour the grounds of the World War II Japanese internment camp in Granada, Colorado.[23] These examples incorporate elements of the ethics of care theory, as they foreground for VR users the emotional experiences of others and foster a feeling of empathy. Rather than VR technology being viewed as a neutral or objective device, these experiences implicate the medium as a collaborator in exposing injustices and engaging users to respond to the embodiments they experience. By combining the embodied simulations of present or past forms of discrimination, bias, and hostility with scenarios that propel people to understand historical or present-day contexts, people are empowered to question or challenge the ideologies of their own lived realities.

Helping people experience and understand the phenomenon of racism is not just a journalistic or advocacy endeavor; it also has become a business. Several VR companies, like Christophe Mallet’s Bodyswaps and Morgan Mercer’s Vantage Point, use immersive experiences presented in VR to offer training to companies and government agencies. Employers can purchase units to allow their employees to occupy the virtual body of someone of another race or gender to reveal their own unconscious bias. The goal of Bodyswaps, as the company describes on its website, is “transforming behaviour, for good.” Bodyswaps explains its experiences as such:

> In BODYSWAPS’ scenarios, professional actors have been motion-captured to give life to virtual humans that talk, move and react like real people. Moreover, the learners get to interact with those virtual humans using their own voice. The combination of social presence and participation unleashes the benefits of hands-on practice usually reserved for face-to-face roleplaying.[24] (See figure 6.2.)

The technological attraction of extending reality and offering immersive experiences is what the company believes will have an impact on those who embrace this technology. Morgan Mercer’s company, Vantage Point, has a similar purpose. It explains on its website, “While technology can cause apathy, immersive technology can drive empathy and fundamentally make the world more human. This is our mantra.”[25] Self-identifying as a survivor of sexual violence, Mercer states that offering anti-sexual harassment training in the workplace is more than a business opportunity. The training is an

Figure 6.2. Bodyswaps VR Workplace Simulation. *Screenshot by author.*

ethical obligation to educate people about the topic and empower survivors by sharing their stories.[26]

Perhaps VR immersive experiences can promote empathy, but is that enough? Ultimately, these programs do offer new opportunities to experience historical injustices in a visceral way, learn from these experiences, and possibly engender a shift in attitudes. However, it is more difficult to prove that VR changes how people think or act. McLean concludes: "However laudable the goals may be, those who engage in cyber-passing still have the luxury of dropping the racial veil whenever they tire of playing with blackness."[27] For other critics, these VR experiences are a form of "identity tourism." Lisa Nakamura writes that the promotion of these experiences is a way to automate racial empathy to massage and alleviate guilt among the technocratic class:

> VR 2.0 titles about refugees and other objects of pathos that situate the viewer outside the point of view of a foreign body invite the user to confuse immersive viewing with access to the actual experience of an-other, to perform co-presence within virtual spaces that can be navigated at will, in contrast to the lack of mobility experienced by the real-life people whose lives are the material for the developers.[28]

Are these experiences a real opportunity to modify thinking and promote change, or are they a form of armchair activism, a way to generate the feeling that users are engaging in action while not really doing anything materially to enact a sustaining form of social change? What impact do the goals of the producers of VR experiences have if the means to express those goals convert the intent of changing how others think into a form of entertainment?

Does the spectacle of VR technology override the social messages in these experiences, or do users internalize these experiences to form lasting change in thinking that may lead to later action?

Scholars of social change emphasize that, for change to be significant and lasting, it must occur both on an individual and group basis as well as on a structural level. In other words, change must be imbued within the core of a society so that the rules and codes that form the basis of social organization can also change. Do immersive experiences motivate people to enact structural change, or are they by their very nature limited to individual experiences of racism that are disconnected from a path to solving structural inequities?

STRUCTURAL RACISM AND VIRTUAL REALITY

In order to address the root causes of structural bias in VR, consideration must be given to examining how the technology incorporates algorithmic bias. Marshall McLuhan often referred to how people receive new technologies as rear-view mirror thinking; that is, people tend to understand new technologies through the lens of previously dominant technologies.[29] For example, television was first described as "radio with pictures" when, in fact, it grew to become a unique medium with its own characteristics and functions that are different from radio. There is a tendency, then, to think of VR as TV in a headset or a video game that is worn on the head, forms of entertainment and distraction that provide an escape from the real world. However, the words "virtual reality" suggest that these experiences are a form of simulated reality, encouraging people to mimic their current environment with its underlying biases, even unconsciously, to provide a modified reality but familiar enough to the one they experience outside of the headset. Likewise, algorithmic biases often reflect the environments in which they are created. Though immersive technologies have the capability to expose the underlying ideologies of social inequities, the developers of these technologies are still very much embedded in these social structures and often derive privilege from them, leading to many experiences in VR that mimic the same biases that exist in real life.

In *Race After Technology*, Ruha Benjamin discusses what she describes as The New Jim Code, which she defines as "the employment of new technologies that reflect and reproduce existing inequities but that are promoted and perceived as more objective or progressive than the discriminatory systems of a previous era."[30] New technologies create the illusion that society has evolved and progressed beyond the prejudices of the past, but they often embed within them the same privilege, racism, and ideology that existed before their invention. In *Algorithms of Oppression*, Safiya Umoja Noble describes the power of algorithms "and the ways those digital decisions reinforce oppressive social relationships and enact new modes of racial profiling, which I have termed *technological redlining*."[31] One concern about the creation and widespread use of algorithmic decision-making is that it is advertised as forward-thinking while being driven by rear-view mirror thinking. Algorithms use data from the past to make decisions about the future. In "Predatory Predictions and the Ethics of Predictive Analytics," Kirsten Martin describes the market incentive for algorithms to inflate the identification of problematic behavior that requires policing. She writes:

> Predictive analytics carries two additional burdens that make the outcome variable more problematic. First, with predictive analytics the outcome is a probability about the future rather than the present. . . . Second, the need to treat people differently based on the value of the outcome variable pushes us to create a dichotomous label in a particularly ambiguous situation.[32]

Given that the technology field is under-represented in minority employment, those leading the decision-making often are training machines from a dominant, white perspective. Because the output of algorithms appears to be mechanical in nature, their decisions are masked to appear neutral.

A viral video illustrates the New Jim Code and the racism embedded in newer technology that Benjamin writes about. In this video, Desi, a Black man, and his white co-worker demonstrate how the face recognition software in a Hewlett-Packard computer would not respond to Desi's face but responded perfectly to his white co-worker's face. Desi ends the video with the proclamation, "I'm going on record and I'm saying it: Hewlett-Packard computers are racist."[33] Desi's video was tinged with humor, but the problem of facial recognition technology can have much more dangerous consequences. Randal Quran Reid was arrested for a crime he did not commit because of poor facial recognition software, resulting in his serving jail time before exculpating evidence that secured his release.[34] The lessons of the Shirley cards have not been learned. When the majority of people who design new technologies are white and are embedded within a culture of underlying social inequities, then the new technologies they invent, like VR, likely will retain the same inequities of the past.

Part of the problem can be found in what critics label the Black Boxes of these algorithms. In this case, Black Box refers to the idea that the systems that compose much of the design behind these new technologies are hidden from view and inspection. The public does not have access to the code, rules, or logic that make up the system. Therefore, most people who use the technology simply trust that the algorithm is accurate enough. Since much of the code is owned by private companies and protected as trade secrets, these companies have no obligation to release their code and have a financial incentive to keep it secret. In their defense, these companies claim that if they were to reveal their underlying code, hackers could corrupt the systems and produce dangerous results, or other companies could copy the code and gain a competitive advantage.

How can the public address the problem of racism embedded within algorithms that make up much of the technology in use today? Some people advocate for a "glass box" approach. Computer scientist Saurabh Bagchi argues in *Scientific American*: "If software is in a glass box, then software testers and well-intentioned hackers can examine it and inform the creators of weaknesses, thereby minimizing cyberattacks."[35] This approach allows for open and transparent communication that can enable the application of the ethics of care, taking into consideration the effects of technology on all different participants within a culture.

Researchers at MIT argue that the public needs to become more AI and VR literate in order to promote the ethical use of these technologies.[36] One analogy that helps to explain this is an automobile. In past decades, most people possessed general knowledge of how to perform minor maintenance and repairs on their car, or at the very least, were aware of the components that made up their car. Now, when most people see a check engine light on their dashboard, they have to go immediately to their car dealer or a mechanic in order to have a computer assess the problem. Even the professional mechanics themselves have to rely on the output of a computer to understand what needs to be repaired or replaced. This has become an accepted practice or an environmental social condition. However, critical assessment must be made of new technologies as they become integrated into more aspects of people's everyday lives, including education, health care, the workforce, and leisure activities. It is worth considering what level of transparency people should demand of the technologies they use and how much they should be allowed to modify them to make them more useful and fair.

Other critics of technology prefer more tangible solutions to this issue. Some scholars argue for what they refer to as Algorithmic Reparations. For example, Jenny L. Davis, Apryl Williams, and Michael W. Yang call for the necessity of actively questioning the design of algorithms: "We intend for this approach to guide algorithmic design and to act as an evaluative standard by which existing algorithmic systems are judged, adjusted, and where necessary, omitted or dismantled."[37] They outline a system to classify and provide restitution to those who have been negatively impacted by the biases embedded in these algorithms. This perspective insists that it is not enough to use virtual technology to make visible existing inequities to people. Instead, a concerted effort must be made to question and actively change the codes behind the technology.

THE VEIL OF IGNORANCE AND RACE IN VR

John Rawls' veil of ignorance is a theory that argues people need to consider facts while shielding themselves from information identifying particular parties or having any personal advantage when making ethical decisions.[38] Rawls points out the deficiency of a utilitarian view of ethics, as the framers of a utilitarian code likely would think of their own self-interest when considering what is defined as a greater good. Rawls concludes, "Thus it seems that the principle of utility is incompatible with the conception of social cooperation among equals for mutual advantage. It appears to be inconsistent with the idea of reciprocity implicit in the notion of a well-ordered society."[39] Instead, Rawls proposes an ethical framework that is constructed without any knowledge of who benefits or is harmed by it. The goal of the theory is to have people remove their own self-interest so that those most vulnerable to a decision are protected. The veil of ignorance benefits marginalized groups by favoring forms of social equity.

Many of the works that use VR to give people the power to embody the experiences of others are produced in the hope that these experiences will modify the implicit biases people have internalized. However, it is not clear how well this works or if VR merely replicates current understandings of race and injustice for people who are aware of these issues. Different people bring different past experiences to VR when they take on the avatar of another racial group. For some, this may be their first exposure to what oppressed groups experience. When experiencing incidents of racism in VR, does that motivate users to work toward the goal of reducing forms of racial injustice, or does the feeling of empathy subside when the HMD is removed?

Conversely, those who belong to oppressed minority groups may want to take on a character belonging to a dominant group to leave behind, even temporarily, the microaggressions that they experience on a regular basis. Do VR experiences like these provide a level of power previously out of reach, or do they merely encourage a feeling of resignation to the prevalent racism in society? VR is largely an individual experience, so responses to immersive experiences vary from person to person. For this reason, an ethics of care model or the veil of ignorance, both of which protect the most vulnerable, may be useful to guide its use. However, for substantial change to happen, these guides must be considered at the stage of production as well as consumption, from headset manufacturers to platform designers to app developers. Unless all of these stages are involved, little structural change is possible.

UTOPIAN AND DYSTOPIAN FUTURES

When a new technology is introduced into a culture, two events usually occur in quick succession. First, people often praise its benefits and potential to help build a better society. During the early days of radio, the medium was thought to become a tool to educate the masses, much like the Internet was first presented as the information superhighway that could revolutionize schools. They saw these technologies as tools to attack injustices and prejudices and reduce bias. Soon afterward, critics warn that the new technology will inflame the worst impulses in humanity and harm civilization. These criticisms are nothing new: Socrates warned about the invention of writing, concerned about its capacity to weaken the memory of men and change how people understood the meaning of knowledge:

> The specific which you have discovered is an aid not to memory, but to reminiscence, and you give your disciples not truth, but only the semblance of truth; they will be hearers of many things and will have learned nothing; they will appear to be omniscient and will generally know nothing.[40]

What would Socrates say about VR, with its ability to simulate alternate realities or generate verisimilitudes of an existing reality? VR appears to be an ideal technology for those who want to "improve" humans by placing them in new environments with different social rules. For others, VR can serve as an escape from hate, biases, racism, or inherent inequities that they face in the physical world. However, users do not leave their own biases behind when they enter virtual worlds, and VR creators may unconsciously, or for some malevolent actors, purposefully perpetuate stereotypes when designing virtual environments and characters. The virtual empathy experienced in VR may, in Socrates' view, be mistaken for the kind of empathy that leads to action to address inequities in the real world.

What is left to be determined is how much producers and consumers of VR should highlight or restrict biases, and what degree of freedom should users have in choosing to be represented differently in VR than in real life. An ethics of care would keep in mind the most vulnerable populations when making these choices, allowing the freedom to experiment with identity and learn from the experiences of others while restricting those who choose to misrepresent people or environments for the purpose of deception that may ridicule or otherwise inflict harm on marginalized groups. With the capacity to reconstruct and distort representations, VR has the power to either perpetuate or challenge existing social biases. The freedom of adopting avatars of different genders, ethnicities, races, or classes comes with the ethical distinction between genuine self-exploration and appropriation or misrepresentation.

Without thoughtfully foregrounding existing biases and prejudices, producers and consumers of VR risk reproducing them in virtual environments.

NOTES

1. Ray-Ban. "Ray-Ban USA Smart Glasses." Accessed September 24, 2023. https://www.ray-ban.com/usa/ray-ban-stories.
2. Noble, Safiya Umoja, and Sarah T. Roberts. "Through Google-Colored Glass(es)." In *Emotions, Technology, and Design*, 187–212. Elsevier, 2016. https://doi.org/10.1016/B978-0-12-801872-9.00009-0.
3. Noble and Roberts, "Through Google-Colored Glass(es)," 9.
4. Apple, Inc. "Apple Vision Pro." *Apple*. Accessed September 24, 2023. https://www.apple.com/apple-vision-pro/.
5. Roth, Lorna. "Looking at Shirley, the Ultimate Norm: Colour Balance, Image Technologies, and Cognitive Equity." *Canadian Journal of Communication* 34, no. 1 (March 30, 2009): 111–136. https://doi.org/10.22230/cjc.2009v34n1a2196, 128.
6. Nowak, Kristine L., and Jesse Fox. "Avatars and Computer-Mediated Communication: A Review of the Definitions, Uses, and Effects of Digital Representations on Communication." *Review of Communication Research* 6 (2018): 30–53. https://doi.org/10.12840/issn.2255-4165.2018.06.01.015, 40.
7. Yang, Grace S., Bryan Gibson, Adam K. Lueke, L. Rowell Huesmann, and Brad J. Bushman. "Effects of Avatar Race in Violent Video Games on Racial Attitudes and Aggression." *Social Psychological and Personality Science* 5, no. 6 (August 2014): 698–704. https://doi.org/10.1177/1948550614528008.
8. Martins, Luis Bravo, and Samantha G. Wolfe. *Metaversed*. Indianapolis: John Wiley and Sons, 2022.
9. Martins and Wolfe, *Metaversed*, 30.
10. Mulenga, Natasha. "Adele on Cultural Appropriation: 'I Didn't Read the F**king Room.'" *Teen Vogue*, October 8, 2021. https://www.teenvogue.com/story/adele-addressed-cultural-appropriation-backlash-bantu-knots-vogue-covers.
11. McLean, Willajeanne F. "Who Are You Wearing? Avatars, Blackface and Commodification of the Other." *Idea* 61, no. 3 (n.d.): 455–506, 485.
12. Baker-White, Emily. "'This Is Blackface': Inside the Virtual Reality Company Trying to Scale Diversity Training." *BuzzFeed News*, December 1, 2021. https://www.buzzfeednews.com/article/emilybakerwhite/diversity-training-mursion-vr-white-actors.
13. Tilley, Elspeth. "New Culture/Old Ethics: What Technological Determinism Can Teach Us about New Media and Public Relations Ethics." In *The Ethics of Emerging Media: Information, Social Norms, and New Media Technology*, edited by Bruce E. Drushel and Kathleen M. German, 191–212. New York: Continuum, 2011, 203.
14. Totilo, Stephen. "VR Project Puts Users in a Different Body to Sense What Gender Dysphoria Might Feel Like." *Axios*, July 27, 2023. https://www.axios.com/2023/07/27/vr-body-of-mine-transgender-experience.

15. D'Olimpio, Laura. "Ethics Explainer: Ethics of Care." *The Ethics Centre* (blog), May 16, 2019. https://ethics.org.au/ethics-explainer-ethics-of-care/.

16. Villegas-Galaviz, Carolina, and Kirsten Martin. "Moral Distance, AI, and the Ethics of Care." *AI & Society*, March 23, 2023. https://doi.org/10.1007/s00146-023-01642-z.

17. McPhate, Mike. "California Today: In Virtual Reality, Investigating the Trayvon Martin Case." *The New York Times*, February 24, 2017. https://www.nytimes.com/2017/02/24/us/california-today-trayvon-martin-case-virtual-reality.html; a short video of the VR experience can be found here: https://www.youtube.com/watch?v=1hW7WcwdnEg.

18. Shipps, Alex. "Simulating Discrimination in Virtual Reality." *MIT News*, January 5, 2023. https://news.mit.edu/2023/simulating-discrimination-virtual-reality-0105.

19. Cogburn, Courtney D., Jeremy Bailenson, Elise Ogle, Tobin Asher, and Teff Nichols. "1000 Cut Journey." In *ACM SIGGRAPH 2018 Virtual, Augmented, and Mixed Reality*, 1–1. Vancouver British Columbia Canada: ACM, 2018. https://doi.org/10.1145/3226552.3226575.

20. Salmanowitz, Natalie. "The Impact of Virtual Reality on Implicit Racial Bias and Mock Legal Decisions." *Journal of Law and the Biosciences* 5, no. 1 (May 1, 2018): 174–203. https://doi.org/10.1093/jlb/lsy005, 181.

21. Salmanowitz, "The Impact of Virtual Reality," 195.

22. Salmanowitz, "The Impact of Virtual Reality," 199–200.

23. Contreras, Russell. "New Virtual Reality Technology Seeks to Teach Racial Empathy." *Axios*, May 15, 2023. https://www.axios.com/2023/05/15/new-vr-role-playing-insight-racism.

24. Bodyswaps. "Bodyswaps: Transforming Behaviours—for Good." Accessed September 21, 2023. https://bodyswaps.co/wp-content/uploads/2021/03/Bodyswaps-Learning-Framework.pdf.

25. Vantage Point. "Vantage Point." Accessed September 25, 2023. https://www.tryvantagepoint.com.

26. Mercer, Morgan. "Vantage Point." Accessed September 25, 2023. https://www.morganmercer.tech/vantage-point.

27. McLean, "Who Are You Wearing?" 490.

28. Nakamura, Lisa. "Feeling Good about Feeling Bad: Virtuous Virtual Reality and the Automation of Racial Empathy." *Journal of Visual Culture* 19, no. 1 (April 2020): 47–64. https://doi.org/10.1177/1470412920906259.

29. McLuhan, Marshall. *Understanding Media: The Extensions of Man*. 1st ed. Cambridge, Ma: MIT Press, 1994.

30. Benjamin, Ruha. *Race after Technology: Abolitionist Tools for the New Jim Code*. Medford, MA: Polity, 2019, 23.

31. Noble, Safiya Umoja. *Algorithms of Oppression: How Search Engines Reinforce Racism*. New York: New York University Press, 2018, 1.

32. Martin, Kirsten. "Predatory Predictions and the Ethics of Predictive Analytics." *Journal of the Association for Information Science and Technology* 74, no. 5 (May 2023): 531–545. https://doi.org/10.1002/asi.24743, 536.

33. Bunz, Mercedes. "Are Hewlett-Packard Computers Really Racist?" *The Guardian*, December 23, 2009. https://www.theguardian.com/media/pda/2009/dec/23/hewlett-packard.

34. Hill, Kashmir, and Ryan Mac. "'Thousands of Dollars for Something I Didn't Do.'" *The New York Times*, March 31, 2023. https://www.nytimes.com/2023/03/31/technology/facial-recognition-false-arrests.html.

35. Bagchi, Saurabh. "Why We Need to See Inside AI's Black Box." *Scientific American*, May 26, 2023. https://www.scientificamerican.com/article/why-we-need-to-see-inside-ais-black-box/.

36. Murray, Scott. "How Artificial Intelligence Can Help Combat Systemic Racism." *MIT News*, March 16, 2022. https://news.mit.edu/2022/how-ai-can-help-combat-systemic-racism-0316.

37. Davis, Jenny L., Apryl Williams, and Michael W. Yang. "Algorithmic Reparation." *Big Data & Society* 8, no. 2 (July 2021): 205395172110448. https://doi.org/10.1177/20539517211044808, 8.

38. Rawls, John. *A Theory of Justice*. Rev. ed. Cambridge, MA: Belknap Press, 2005.

39. Rawls, *A Theory of Justice,* 13.

40. Plato. "Phaedrus." Translated by Benjamin Jowett, 2008. https://www.gutenberg.org/files/1636/1636-h/1636-h.htm.

Chapter 7

Harassment and Abuse in VR

One of the first VR experiences showcased at the Whitney Biennial exhibition in New York City demonstrated the technology's capacity to depict violence. Jordan Wolfson is an artist known to implicate viewers in the objectification of violence: his 2015 kinetic installation *Colored Sculpture* consisted of a large marionette suspended by steel chains, which alternatively lifted the puppet up into the air and released it to crash back to the ground. In his 2017 exhibition at the Whitney Biennial, Wolfson debuted Real Violence, a passive VR animation that showed a figure on a street corner beating someone else on the sidewalk. Wolfson's own voice recited Hebrew prayers into viewers' ears while they witnessed the figure kicking, stomping, and hitting with a bat the person lying down on the ground. To make the violence seem more real, Wolfson recorded himself beating a life-sized doll to use as a template for the experience.

Wolfson claimed the project attempts to capture the visceral feeling of watching violent videos online,[1] but several art critics wrote about how the experience inflicted its own form of violence on viewers. Nate Freeman noted that the experience implicated viewers in bearing witness to the act. He writes about the piece, "It was virtual, but the beatdown was real."[2] Writing for the *New Yorker*, Alexandra Schwartz argued that the context of witnessing violence helps to reframe the responsibility of viewing: "If Wolfson's video were a documentary, there would be no excuse for what it shows. If it were a scripted movie, with Wolfson slotted into the hero's role, we'd cheer for him from the first crushing skull crack."[3]

The Whitney managed long lines of patrons waiting to experience Real Violence. The attraction of VR as spectacle contributed to its ability to generate attention, but experiencing a piece that promised to shock viewers and prompt them to question their moral judgment by choosing to participate in the experience also played a part. Several viewers I spoke to at the exhibition said that this was the first time they had worn a VR headset. Others had worked with VR before but had read about the violence depicted in this immersive medium. As I have written elsewhere, this exhibit implicates viewers in their choice to witness what is depicted.[4] Viewers could always remove the VR headset at any time to exit the experience, but in the 30 minutes I observed viewers, no one had done so. (See figure 7.1.)

The desire to seek out unsafe experiences or witness violence from behind the safety of a VR headset echoes experiences with older media like violent movies or television shows. However, the perceptual immersion that engages more senses than these older media makes VR a technology that can generate more powerful emotional reactions. Within the context of a consensual relationship between viewers and producers to engage in such an experience, violent VR can prove to be entertaining, cathartic, or even profound. Becoming a virtual witness to violence like that shown in Wolfson's installation may even lead viewers to question the moral stance of bystanders paralyzed by similar acts in the real world.

Wolfson's art demonstrates how powerful VR can be in conveying a perceptual sense of violence, but it is another matter entirely when that violence is enacted against someone who never had sought out that experience. Social VR has become a medium where incidents of harassment or abuse are prevalent. According to a report from the Center for Countering Digital Hate, researchers using the VR Chat feature in Facebook in 2021 encountered an incident of abuse and harassment an average of every seven

Figure 7.1. Patrons Experience Real Violence. *Photo by author.*

minutes.[5] Anne Maass, Mara Cadinu, and Silvia Galdi explain that harassers are often motivated by a desire to exert power or reinforce a sense of identity that belongs to a dominant group.[6] Given the heightened sense of copresence because of its sensory immersive capability, VR opens up opportunities for harassers to engage in behavior that has an outsized effect compared to other online environments. Conducting in-depth interviews with 30 users of social VR, researchers found that harassment is often targeted toward marginalized groups, such as members of the LGBTQ community, women, and those belonging to targeted ethnicities or countries of origin. This form of harassment differs from other digital media because the offensive acts often simulate physical harassment, like unwanted touching or groping. These researchers label this form of abuse "embodied harassment" to address the sense that virtual bodies are representative of a user's physical presence in the virtual environment.[7] A survey of social VR users revealed that "49% of women have experienced sexual harassment in social VR, 30% of men were subject to homophobic or racist remarks, and 20% of men reported violent comments or threats."[8]

If bias is a predisposition to beliefs, harassment weaponizes those beliefs into action. Users can experiment with antisocial behavior while interacting with automated intelligent agents, otherwise known as NPCs. However, when they engage in the same behavior with other real people interacting within the same immersive environment, this behavior has real-world consequences. Like other interactive media, such as social media and multiplayer video games, VR has the same capacity for harassing and abusive communication, though its immersive perception can intensify these acts to the point of blurring virtual and physical experiences. This difference in degree that is unique to VR is important to examine, but it is helpful to first examine how people in older digital environments have contended with virtual abuse.

HARASSMENT AND ABUSE IN DIGITAL ENVIRONMENTS

Long before the immersive capability of VR was widely available, some networked personal computer users constructed text-based virtual environments called MUDs (Multiuser Dungeons) for online adventure gameplay. MUDs were influenced by the role-playing game Dungeons and Dragons, where a dungeon master guides players through an imaginary environment as they seek out a quest or treasure. First running on research computers and electronic bulletin board services where users would dial a phone number to connect their personal computer to a server, MUDs quickly grew as players gained access to the Internet. These largely text-based systems allowed users to set up fantasy environments where they could take on different personas and engage in role-play adventure-based games. Because characters were identified by handles and environments were described in text, MUDs became a popular way of experimenting with identity while preserving anonymity.[9]

In 1992, a character named Mr. Bungle entered the virtual living room of the MUD named LambdaMOO and virtually assaulted several other characters. Using a programming trick known as a voodoo doll, Bungle forced other characters to perform sexual acts on themselves before he was booted from the space by a wizard. Three days later, users gathered in the MUD to debate what to do about the offending character and how to set up and enforce rules of behavior for the virtual space.[10]

During the debate, players argued about the seriousness of the act: Was the text-based offense a serious crime or protected under free speech? Many agreed that the symbolic acts, while not actually physical, had a real psychological effect that resembled what is experienced during a physical assault. A player with the username HerkieCosmo offered that, "In MOO, the body IS the mind."[11] *The Village Voice* editor, Julian Dibbell, reflected on the reality of experience in virtual spaces. He writes, "What happens inside a MUD-made world is neither exactly real nor exactly make-believe, but profoundly, compellingly, and emotionally meaningful."[12] The character Mr. Bungle was deleted from the MUD, but Dibbell noted that a similar character later signed on that seemed to be controlled by the same person in a sort of virtual resurrection. Like with video games, virtual environments can allow users to experience more than one life, just as they can create multiple identities. No punishment is permanent, making the ability to regulate behavior even more difficult.

Two decades later, social media platforms created an environment that provided new opportunities for harassment and abuse. According to the Pew Research Center report on online harassment, over 40% of Americans experienced some form of online harassment in 2020, with the number of Americans experiencing severe forms of harassment growing since 2014.[13] Depending on their design, platforms alternatively restrict users to use a real identity, allow them to use pseudonyms, or be completely anonymous. The ability to hide identities behind anonymous usernames contributes to the problem, as harassers often can escape consequences for sending abusive messages.

However, restricting users to using real identities within social media platforms paradoxically may allow for a form of harassment in real life. Connecting online representations to real people has helped to give rise to the phenomenon of doxing, or publicly posting the personal information of users, including their addresses, phone numbers, identification numbers, places of employment, and information about family members, medical conditions, and finances. While online groups have used doxing in the past to expose people behind anonymous posts that contain illegal activity or socially regressive content, more recently it has been employed for the purposes of stalking, harassment, and financial crimes. A 2022 survey conducted by the Anti-Defamation League reports that, while 5% of Americans have been doxed, 60% of those Americans revealed that the doxing had occurred within the last 12 months. The survey notes that the majority of victims of online harassment belong to marginalized groups, such as ethnic or religious minoritized groups and members of LGBTQ+ communities.[14]

One of the most infamous cases of doxing occurred in what has become known as Gamergate. In 2013, independent video game designer Zoë Quinn launched the game Depression Quest to raise awareness about mental illness. After Quinn received favorable reviews from video game critics, their former boyfriend Eron Gjoni published a blog post that falsely accused Quinn of using sexual favors with journalists so that they would write favorable reviews of their game. Anonymous users shared the post and began a campaign against Quinn, other feminist game designers, and their supporters. This campaign included death threats and publicly posting home addresses of the victims. Quinn was forced to leave their home out of fear of violence.[15] A study of users claiming to have participated in Gamergate reported that 89.2% were male, 74.5% were white, and 74.1% identified as heterosexual.[16]

University of New South Wales professor Michael Salter argues that Gamergate was assisted through the technological rationalization of the computing industry that generated a masculine subjectivity throughout it. Technological systems are not neutral tools, according to Salter. Instead, they imbue the values and biases of the culture in which they are built, and their inherent design is made to conform to those values. The videogame industry's embrace of male aggression in the form of "geek masculinity" serves to repel anyone who does not identify with these values. He writes:

> Without the underlying misogyny of geek masculinity and the technological rationality that sustains it, Gjoni's article would likely have been another obscure, if hurtful, blog post. It was the lifeworld of geek masculinity and its correspondence with online technological architecture and administration that formed the condition of possibility for the abuse campaign that followed.[17]

In effect, Gjoni's blog post was the spark that initiated the crowdsourcing efforts of thousands of users who enforced a white, straight, male hegemony by threatening the lives of those who challenged it. While Gjoni had distanced himself from those making threats of violence, anonymous online forums encouraged users to dox Quinn and other feminist gamers, send messages threatening rape and other forms of physical violence, and make SWATing calls, where users call police and falsely report they are experiencing a crime at a victim's address, causing an emergency police response to that location.

Since these events, Quinn co-founded the Crash Override Network, an organization that provides resources to victims of online harassment and abuse, as well as offering advice and providing referrals to law enforcement about cases of harassment. Crash Override had also featured a crisis phone hotline to support victims but suspended that service in 2016 because the number of cases had overwhelmed the organization.[18] Other organizations have since emerged: Online SOS, which provides several checklists for people to follow if they become victims of doxing, online threats of violence, mob harassment, or distribution of false information or private photos.[19] The Coalition Against Online Violence, made up of academic, nonprofit, and industry partners, offers an online violence resource hub that includes legal and psychological assistance for journalists who are doxed, hacked, or threatened.[20]

ETHICS OF DIGITAL INFORMATION

These organizations seek to protect and support victims of online harassment and abuse, which is clearly an ethical response to this phenomenon. However, are there any ethical frameworks in which doxing may be permitted? While a deontological theory of virtual ethics supposedly would oppose doxing in any case as a universal standard, a libertarian ethics that declares all information should be free and available negates the possibility of privacy online. While this ethic, based on the concept of individual liberty, rejects the ability of state actors to gather and use information as a means of control, it also rejects the prohibition of individuals from accessing and using this information within a free market.[21] Valuing systematic thinking over a sense of fairness or protection of vulnerable groups, a libertarian ethics edges toward a deontological position that doxing should not be regulated by a political stance or fear of consequences. For example, the civil libertarian Alan Dershowitz argues that doxing should not be curbed based on a political metric or concern about the possible effects on victims, as the right to publish information is protected speech under the First Amendment. Acts of violence and threats should be prosecuted, according to Dershowitz, but not the act of sharing information, even if it is personally identifiable.[22]

In contrast, a utilitarian ethic may evaluate the effect of doxing as a measure of the number of people who are helped or harmed by the act. While doxing in cases like Gamergate is widely understood as unethical conduct, some activists have argued that doxing can be a useful and ethically justified action against those who use anonymity as a shield while they engage in fear and intimidation of marginalized groups. For example, Michael Cullinane writes for the Center for Digital Ethics and Policy that it is ethical to dox self-proclaimed Nazis if done responsibly. After the 2017 Unite the Right rally in Charlottesville, Virginia, many people crowdsourced information to identify attendees and publish their names, addresses, places of work, and other identifiable information. Cullinane suggests a justification for doxing those attending the rally as follows: "If the Nazis are afraid of being publically [*sic*] shamed, they're less likely to be aggressive in their public and online behavior."[23] Likewise, Peter Brian Berry argues that doxing of sexual transgressors as a part of the #MeToo movement is justified when normal legal channels to serve justice have failed.[24]

While considering how to apply an ethical framework to the problem of online harassment and abuse, even the study of these phenomena comes with their own ethical risks. A group of scholars has suggested that researchers should apply an ethics of care when examining the online users who participated in doxing as a part of Gamergate. With implications across the digital humanities, these scholars argue that deontological ethics are often employed by institutional review boards through the establishment of clear guidelines about what types of research and researcher conduct are ethical, and utilitarian ethics within the area of biomedical research, where the risks of a potential treatment should be weighed against the good that could come from discovering an effective treatment or cure for a disease. However, an ethics of care prioritizes the relationship between a study's researcher and participant so that the rights of the participant are respected within the asymmetrical power relationship within the study environment. For example, repeating some of the messages and threats spread during the scandal risks further traumatizing victims or inflicting trauma on research assistants engaged in the work while helping doxers spread the harassing messages.[25]

When ethical guidelines gain enough political support to be enforced, they often are codified into law. However, at least in countries with non-authoritarian governments, laws take time to be established, properly enforced, and adjudicated. In the meantime, digital information and communication systems evolve so quickly that laws cannot keep up with these changes. When Elon Musk bought Twitter and renamed the company X, he also sought ways to raise revenue for the company. One idea that he implemented was to charge users for the right to display a blue checkmark to verify their identity, opening the floodgates for people to impersonate others simply by paying X a monthly fee. Boston University electrical and computer engineering professor Gianluca Stringhini described the problem with this change:

> Now anyone willing to pay $8 can obtain a check mark and impersonate any entity. This opened up opportunities for any type of scam, and we saw many

> examples of this. . . . Before this change, a malicious actor would have had to compromise a real, verified account—for example by stealing their password—and send a tweet through that account, by leveraging their [the real account owner's] reputation.[26]

U.S. representative Alexandria Ocasio-Cortez wrote on X that the change made the system "ripe for targeted harassment, especially of women + others on this platform."[27]

Ironically, the law can remedy the problem of impersonation more so when it comes to products instead of people. In the virtual community Second Life, owner Linden Lab informed the user Frolic Mills that their use of the phrase Miss SL Universe prompted a complaint from Miss Universe L.P. that its trademark was being violated. As a result, Linden Lab unilaterally removed any reference of Miss Universe and Miss SL Universe from the user's profile, and it requested that the user refrain from using those names in the future.[28] Commenting on a court case against another Second Life user that copied and sold virtual poses and products within the platform, intellectual property attorney Jorge Contreras Jr. stated, "Virtually every aspect of real life is getting duplicated, and all the laws that can be applied to the real world are being applied in 'Second Life.'"[29] Legal issues in the virtual environment spawned the Second Life Bar Association, a group that maintains an office on the virtual world Justitia and had hosted a discussion board about legal issues like Fair Use on the Internet, declaring war in a virtual world, and seeking legal redress against harassment or abuse when attacks occur across several jurisdictions.[30]

Beyond virtual environments themselves, predictive and generative AI algorithms are technologies that are quickly changing how these environments function and respond to human engagement. The rapid adoption of consumer-facing AI since the public release of OpenAI's ChatGPT in November 2022 has sparked debate about placing limits on the development and use of digital media. The ability of AI tools to generate deepfake videos, or video and audio of people saying words or doing actions that never actually occurred in real life, has prompted calls for new regulations and laws. A bipartisan group of U.S. senators has proposed the NO FAKES Act, prohibiting any AI-produced video using a digital replica of a person without permission.[31] Warning that "Artificial Intelligence is making it easier to extract, re-identify, link, infer, and act on sensitive information about people's identities, locations, habits, and desires," President Biden issued an executive order in October 2023 that directs federal agencies to protect against the "improper collection and use of people's data,"[32] and the European Union has passed a comprehensive framework of regulations to mitigate against cognitive behavioral manipulation, biometric identification, and use of generated content to misinform people or generate deepfakes.[33] All of these proposals will have an impact on the use of AI to generate virtual environments in which people interact with these systems and with one another.

ETHICS OF ABUSE IN AND OF VR

A problem with current law is that it is not set up to distinguish between speech and actions in the real world as opposed to virtual ones. Some law scholars argue for the need to pass laws that regulate virtual worlds as they grow and become more of a simulation of our real world. One difference between the physical and virtual worlds is that, in the physical world, no two people or objects can occupy the same space, making location a limited and valuable commodity. In the virtual one, a potentially infinite stretch of virtual space can be created, as well as an infinite number of copies of the same space. One possible solution for dealing with disputes is to set up multiple worlds, each with its own set of rules. Discussing the management of legal disputes in Second Life, one law review article suggests:

> For the business that wants to test out a real-world business plan, they can join the world where players retain intellectual property rights, and everyone in the world is aware of those rights. For the weekend player that is not concerned with retaining rights to the outfit she created, she can join the world where Linden Labs [*sic*] holds all intellectual property rights.[34]

Setting up multiple separate virtual worlds with different rules to appeal to each individual as a way of preventing conflict may appear to solve the problem of virtual abuse, but this solution does not address the real communication that takes place within these worlds. On the level of symbolic representation, the relative distance between virtual avatars and objects is real. In fact, one valuable aspect of a virtual world is in its representations of the people who make up that world. When a real person is virtually controlling an avatar, the communication that occurs is different than when the message is generated by an NPC. In this way, a shared virtual environment performs a function similar to a telephone or video chat more so than a video game or fictional narrative.

Like people sending threatening messages during the Gamergate controversy, VR allows harassers to hide behind avatars as they inflict harm. Andrew Bosworth, Chief Technology Officer of Meta, warned employees in a memo cited in the *Financial Times* that women and minoritized groups often face a "toxic environment" in VR, where automatic moderation of virtual spaces would be difficult to scale.[35] While Bosworth suggests that the same moderation practices used for Meta's other platforms could be used in VR, he acknowledged that the "immersive nature of virtual reality" could enhance the negative effects of abusive behavior.[36] With the adoption of Meta's social media policy, moderators could provide warnings to users who engage in abusive behavior and suspend their accounts for repeat offenses.

The problem of only reacting to abuse once it occurs, however, means that initial cases of abuse and harassment would still be able to take place. The principle of no prior restraint is an outgrowth of the First Amendment, but when no prior restraint is coupled with the power of crowdsourcing, the result can be a deluge of online attacks, hacking, and real-life effects like SWATing and physical attacks. Kavya Pearlman from the XR Safety Institute expresses concerns about the self-censorship of VR users, particularly of vulnerable groups, out of fear of being singled out and harassed on immersive platforms. Regarding building safety precautions to guard against such attacks, Pearlman asks, "What kind of reality do we want to create?"[37] In order to prevent being harassed, vulnerable users may choose to hide or masquerade as white, cisgender male avatars, further closing off the presentation of diverse representations of these spaces and implying that the metaverse is an environment primarily for white men.

For Kashmir Hill, *The New York Times* reporter whose personal space was virtually violated in Meta's Horizon Worlds space, the only recourse available to her was to either leave that space or call for a vote of participants in that space on whether to remove the offending avatar.[38] These options would require alternatively disengaging completely or calling attention to the action and disrupting the activity occurring before the harassment. While Hill found the experience annoying, others who have been subjected to abuse in VR have described the immersive experience as traumatizing. One of the more well-known cases is described by Jordan Belamire, who played a multiplayer VR game on an HTC Vive HMD while her husband and brother-in-law watched on a nearby screen. As she played, another user with the tag BigBro442 began groping and chasing her, placing his virtual hands on the body of her avatar. Despite her first attempts to laugh off the encounter, Belamire recognized the event as feeling like a violation of her body. She writes,

> Remember that little digression I told you about how the hundred-foot drop looked so convincing? Yeah. Guess what. The virtual groping feels just as real. Of course, you're not physically being touched, just like you're not actually one hundred feet off the ground, but it's still scary as hell.[39]

As virtual environments become more widespread and are presented alternatively as fun leisure spaces, learning institutions, or virtual places of work, the opportunities for harassment will continue to increase unless proper safeguards are designed, implemented, and enforced. Recently, police in the United Kingdom were called to investigate what has been determined to be the first virtual sexual assault investigation in the country. A girl under 16 reported that her avatar was attacked online, which caused her psychological harm despite not being physically threatened. A BBC article reporting the case stated, "According to an unnamed senior officer familiar with the matter who spoke to the paper the victim, under 16 at the time, suffered psychological trauma 'similar to that of someone who has been physically raped.'"[40]

The emerging rules of social virtual environments demonstrate that there are no clear boundaries that separate the physical world from VR. A Pew Research Center study found that a majority of experts in technology fields believed that the metaverse will be a part of daily life for a large number of people by 2040, but that experience will exist in differing layers depending on the context of use, from a digital tool that helps productivity to a leisure activity to making social connections to providing access to education and more. Some may jump in and out of VR to complete specific tasks, while others could spend many of their waking hours alone or interacting with others in the metaverse for work and play.[41] Each of these contexts can generate what Benedict Anderson has called an "imagined community," a formulation that Anderson applied to the conception of nationalism but can easily represent virtual communities. Anderson writes, "Communities are to be distinguished, not by their falsity/genuineness, but by the style in which they are imagined."[42] A social construction of virtual spaces requires a shared understanding of what constitutes membership in the community, the cultural values and rules of that space, and the means of enforcing these rules and protecting against threats from outsiders.

One difference between imagined communities in the physical and virtual worlds is the malleability of virtual space. When the structure and membership of any one virtual space can change so quickly, it is difficult to preserve a stable foundation to establish and preserve a consistent set of social rules and understood boundaries. The relative lack of consequences from misbehavior, as well as a sense of performance as users control avatars in VR, disadvantages the shared sense of community even as users interact with one another in real time.[43] However, as Byron Reeves and Clifford Nass have argued in *The Media Equation*, people often treat their interactions with machines as if they were real people. Because technology has evolved faster than biological evolution, people often tend to anthropomorphize technologies with which they interact and respond to them as they would with humans in social contexts when machines mimic human behavior or physical characteristics.[44] Within VR spaces, people sometimes treat other people as fictional characters that they can manipulate as if they were playing a game, while they also respond to digital objects as if they were human.

Given that the shifting social conventions of VR create such an unstable foundation, the American Bar Association is debating how to apply current law to virtual worlds. One essential question that has been raised is whether the owners of immersive spaces are responsible for the actions between users within them. Owners conceivably could generate code that restricts the expression or actions of users in order to guard against liability; however, that raises the issue: "whether the environment and user communication in virtual public spaces will have the same First Amendment freedom as newspaper or other entertainment mediums."[45] Another issue is the corporality of crimes, as in the case of assault or homicide: "The issue here is as significant as it is simple: tort law generally regulates the physical world, not the virtual. In a virtual world, there are not grounds for battery or assault because there is no actual physical contact."[46] If an avatar is an extension of a person's personality into the virtual environment, then does the deletion of a user's account, either because of a violation of a platform's rules or as a result of nonpayment of a subscription that provides access, constitute a sort of execution that causes psychological harm to a person? People in many countries have the right of due process in the physical world. In virtual spaces that viscerally feel real, however, these rights may carry different weight, especially if an owner argues that the space is not real but merely code. Users themselves often report that, on an intellectual level, they recognize that what they are engaged in is not necessarily "real." Studies have shown, though, that exposure to virtual "bad guys" increases users' stress levels and generates physical responses that can harm users.[47]

Although the examples presented throughout this chapter suggest that regulation of VR is needed to protect users from incidents of abuse and harassment, laws and policies that control expression and behavior can have the effect of becoming too restrictive. Carlos Gutierrez, Deputy Director and General Counsel for the LGBT Technology Partnership and Institute, expresses concern that an ethical framework for VR that restricts sexual appearance and behavior to heteronormative representations may suppress inclusion of the LGBTQ+ members. Developers "should keep in mind design decisions that may ostracize potential users, such as LGBTQ+ members," Gutierrez advises. "This can start with offering customization options for avatars and non-heteronormative activities."[48] Sexual harassment training that uses VR simulations, such as Vantage Point and Sisu VR, can provide users with immersive experiences that show how harassment appears from the point of view of victims in order to nudge people toward more ethical behavior.[49] However, there is concern that those who face harassment and degrading treatment at work will be retraumatized by mandatory VR training that presents similar incidents in an immersive and emotionally intense way.[50] Tensions between safety and freedom require constant renegotiation of policies that allow people to freely express themselves while protecting them from harassing behavior from others or threats to their psychological and physical safety.

One seemingly successful way to regulate behavior within imagined VR communities is the implementation of user reviews. One study had found that a properly implemented reviewing system works better than a platform-based regulation of behavior. Analyzing one thousand user reviews, researchers identified over one hundred posts that reported either direct experience of or the witnessing of an incident of harassment. Capturing not only the number of incidents but also the nature of each case, a qualitative approach to user reviews can help to identify patterns and understand how to mitigate abuse.[51] Such a system must be monitored so that the review process is not gamed by crowdsourcing positive or negative reviews to achieve the outcome of keeping an abusive user on a platform or removing a victim from it.

Ethically producing individual VR experiences should involve disclosure of what someone will experience so that they are not subjected to unwanted content. However, VR quickly is becoming more of a social experience in which people interact with a range of pre-programmed NPCs, AI-adaptive agents, and real people manipulating increasingly photorealistic avatars. Real people bring their own experiences and habits into social VR, including harassing and abusive behavior. While antisocial and psychopathic behaviors have existed in the physical world long before the invention of VR, the unreality of digital representation combined with pseudonymity or anonymity provides more opportunities for offending behavior to occur unchecked. Harassment, often directed toward vulnerable groups, can have the effect of driving members of those groups away from social VR, making those platforms less diverse and less welcoming to those groups.

While VR may be assumed to not be a part of the physical world proper, its immersive nature often feels real, and real people communicate with one another within social VR. As with individual VR experiences, social VR should be bound by a social contract between users that guarantees the prohibition of unwanted behavior. Rather than banning certain expressions or actions outright, a social contract approach sets the ground rules to allow interactions to take place between users only if both parties agree. An ethics of care model can help keep vulnerable groups in mind when designing systems so that those without a voice are not harmed by others who control the conversation. The boundaries between the unrealness of VR and the physical, financial, and psychological consequences outside of VR have blurred. Unreal violence is real if it is unwanted, and real policies must protect those subjected to it.

NOTES

1. Kaplan, Isaac. "Jordan Wolfson's VR Work, 'Real Violence,' Has the Art World Talking." *Artsy*, March 17, 2017. https://www.artsy.net/article/artsy-editorial-gut-wrenching-vr-work-art-talking-violence.

2. Freeman, Nate. "A History of Violence: Jordan Wolfson on His Shocking Foray into VR at the Whitney Biennial." *ARTnews.Com* (blog), March 1, 2017. https://www.artnews.com/art-news/artists/a-history-of-violence-jordan-wolfson-on-his-shocking-foray-into-vr-at-the-whitney-biennial-7856/.

3. Schwartz, Alexandra. "Confronting the 'Shocking' Virtual-Reality Artwork at the Whitney Biennial." *The New Yorker*, March 20, 2017. https://www.newyorker.com/culture/cultural-comment/confronting-the-shocking-virtual-reality-artwork-at-the-whitney-biennial.

4. Grabowski, Michael. "Perception and Poetics of VR Documentaries." Edited by Christian Iseli. *zDok.17: Being There*, November 2017, 1–12. https://blog.zhdk.ch/zdok/files/2017/11/referat_grabowski.pdf.

5. Center for Countering Digital Hate. "Facebook's Metaverse." *CCDH*, December 30, 2021. https://counterhate.com/research/facebooks-metaverse/.

6. Maass, Anne, Mara Cadinu, and Silvia Galdi. "Sexual Harassment: Motivations and Consequences." In *The SAGE Handbook of Gender and Psychology*, edited by Michelle K. Ryan and Nyla R. Branscombe, 341–358. 1 Oliver's Yard, 55 City Road London EC1Y 1SP: SAGE Publications, Ltd, 2013. https://doi.org/10.4135/9781446269930.n21.

7. Freeman, Guo, Samaneh Zamanifard, Divine Maloney, and Dane Acena. "Disturbing the Peace: Experiencing and Mitigating Emerging Harassment in Social Virtual Reality." *Proceedings of the ACM on Human-Computer Interaction* 6, no. CSCW1 (March 30, 2022): 1–30. https://doi.org/10.1145/3512932.

8. Outlaw, Jessica. "Virtual Harassment: The Social Experience of 600+ Regular Virtual Reality (VR) Users." *The Extended Mind*, April 4, 2018. https://www.extendedmind.io/the-extended-mind-blog/2018/04/04/2018-4-4-virtual-harassment-the-social-experience-of-600-regular-virtual-reality-vrusers.

9. MIT professor Sherry Turkle documented the culture of MUDs and the early Internet in: Turkle, Sherry. *Life on the Screen: Identity in the Age of the Internet*. New York: Simon & Schuster, 1995.

10. Dibbell, Julian. "A Rape in Cyberspace: How an Evil Clown, a Haitian Trickster Spirit, Two Wizards, and a Cast of Dozens Turned a Database into a Society." *The Village Voice*, December 23, 1993. https://www.villagevoice.com/a-rape-in-cyberspace/.

11. Dibbell, "A Rape in Cyberspace."

12. Dibbell, "A Rape in Cyberspace."

13. Vogels, Emily A. "The State of Online Harassment." *Pew Research Center: Internet, Science & Tech* (blog), January 13, 2021. https://www.pewresearch.org/internet/2021/01/13/the-state-of-online-harassment/.

14. Anti-Defamation League Center for Technology & Society. "Online Hate and Harassment: The American Experience 2022," 2022. https://www.adl.org/sites/default/files/pdfs/2022-06/Online-Hate-and-Harassment-Survey-2022-v7.pdf.

15. Quinn, Zoë. *Crash Override: How Gamergate (Nearly) Destroyed My Life, and How We Can Win the Fight against Online Hate*. New York: PublicAffairs, 2017.

16. Ferguson, Christopher J., and Brad Glasgow. "Who Are GamerGate? A Descriptive Study of Individuals Involved in the GamerGate Controversy." *Psychology of Popular Media* 10, no. 2 (April 2021): 243–247. https://doi.org/10.1037/ppm0000280.

17. Salter, Michael. "From Geek Masculinity to Gamergate: The Technological Rationality of Online Abuse." *Crime, Media, Culture: An International Journal* 14, no. 2 (August 2018): 247–264, 259. https://doi.org/10.1177/1741659017690893.

18. "About the Network." *Crash Override Network*. Accessed May 15, 2024. http://www.crashoverridenetwork.com/about.html.

19. Online SOS can be reached at https://www.onlinesos.org.

20. Key partners of the coalition include The Knight Foundation, the Emmerson Collective, and the Craig Newmark Philanthropic Fund. https://onlineviolenceresponsehub.org.

21. Iyer, Ravi, Spassena Koleva, Jesse Graham, Peter Ditto, and Jonathan Haidt. "Understanding Libertarian Morality: The Psychological Dispositions of Self-Identified Libertarians." Edited by Liane Young. *PLoS ONE* 7, no. 8 (August 21, 2012): e42366. https://doi.org/10.1371/journal.pone.0042366.

22. Alan Dershowitz. "When Is Doxxing Ethical or Unethical?" *The Hill* (blog), December 11, 2023. https://thehill.com/opinion/judiciary/4353133-when-is-doxxing-ethical-or-unethical/.

23. Cullinane, Michael. "The Ethics of Doxing Nazis." *Center for Digital Ethics & Society*, October 12, 2017. https://digitalethics.org/essays/ethics-doxing-nazis.

24. Barry, Peter Brian. "#MeToo and the Ethics of Doxing Sexual Transgressors." In *The Palgrave Handbook of Sexual Ethics*, edited by David Boonin, 507–523. Cham: Springer International Publishing, 2022. https://doi.org/10.1007/978-3-030-87786-6_30.

25. Suomela, Todd, Florence Chee, Bettina Berendt, and Geoffrey Rockwell. "Applying an Ethics of Care to Internet Research: Gamergate and Digital Humanities." *Digital Studies/Le Champ Numérique* 9, no. 1 (February 13, 2019): 4. https://doi.org/10.16995/dscn.302.

26. Barlow, Rich. "Does Twitter, under Elon Musk, Need Government Regulation?" *BU Today*, November 18, 2022. https://www.bu.edu/articles/2022/do-elon-musk-and-twitter-need-government-regulation/.

27. Dean, Grace. "AOC Says Elon Musk's Changes to Twitter's Blue Check Feature Are 'Ripe for Targeted Harassment.'" *Business Insider*, November 10, 2022. https://www.businessinsider.com/elon-musk-aoc-twitter-blue-check-notification-harassment-verified-subscription-2022-11.

28. Vendetta, Twisted. "Your Second Life Could Be Changing Soon (Alternate Title: The Clothes Can Stay on the Rack for the Day)." *The Twisted Perspective* (blog), October 23, 2008. https://twistedandtawdry.wordpress.com/2008/10/23/your-second-life-could-be-changing-soon-alternate-title-the-clothes-can-stay-on-the-rack-for-the-day/.

29. Davis, Phil. "Virtual Sex Software Spawns Lawsuit." *NBC News*, August 10, 2007. https://www.nbcnews.com/id/wbna20214184.

30. "SL Bar Association." Accessed January 8, 2024. http://slbarassn.ning.com/.

31. Kenedy, Jennifer, and Jorden Rutledge. "The NO FAKES Act: With Proposed Bill, Congress Set to Protect Against Unauthorized Digital Replicas of Faces, Names and Voices." *JD Supra*, October 17, 2023. https://www.jdsupra.com/legalnews/the-no-fakes-act-with-proposed-bill-2285269/.

32. The White House. "Executive Order on the Safe, Secure, and Trustworthy Development and Use of Artificial Intelligence." October 30, 2023. https://www.whitehouse.gov/briefing-room/presidential-actions/2023/10/30/executive-order-on-the-safe-secure-and-trustworthy-development-and-use-of-artificial-intelligence/.

33. European Parliament. "EU AI Act: First Regulation on Artificial Intelligence." August 6, 2023. https://www.europarl.europa.eu/news/en/headlines/society/20230601STO93804/eu-ai-act-first-regulation-on-artificial-intelligence; European Parliament. "Artificial Intelligence Act: MEPs Adopt Landmark Law." March 13, 2024. https://www.europarl.europa.eu/news/en/press-room/20240308IPR19015/artificial-intelligence-act-meps-adopt-landmark-law.

34. Tricker, Brandy. "Taming the Wild West: Solving Virtual World Disputes Using Non-Virtual Law." *Rutgers Computer and Technology Law Journal* 35, no. 1 (September 22, 2008). https://www.thefreelibrary.com/Taming+the+Wild+West%3A+solving+virtual+world+disputes+using+non-virtual...-a0207060728.

35. Murphy, Hannah. "How Will Facebook Keep Its Metaverse Safe for Users?" *Financial Times*, November 12, 2021. https://www.ft.com/content/d72145b7-5e44-446a-819c-51d67c5471cf.

36. Murphy, "How Will Facebook Keep Its Metaverse Safe for Users?"

37. DEV Studio. *Kavya Pearlman Guest Lecture: Past, Present, and Future of XR—XR Safety Initiative*, February 13, 2021. https://www.youtube.com/watch?v=Omi64iUqTPE.

38. Hill, Kashmir. "This Is Life in the Metaverse." *The New York Times*, October 7, 2022. https://www.nytimes.com/2022/10/07/technology/metaverse-facebook-horizon-worlds.html.

39. Belamire, Jordan. "My First Virtual Reality Groping." *Mic*, October 21, 2016. https://www.mic.com/articles/157415/my-first-virtual-reality-groping-sexual-assault-in-vr-harassment-in-tech-jordan-belamire.

40. Vallance, Chris. "Police Investigate Virtual Sex Assault on Girl's Avatar." *BBC News*, January 2, 2024. https://www.bbc.com/news/technology-67865327.

41. Anderson, Janna, and Lee Rainie. "The Metaverse in 2040." *Pew Research Center*, June 30, 2022. https://www.pewresearch.org/internet/2022/06/30/the-metaverse-in-2040/.

42. Anderson, Benedict. *Imagined Communities: Reflections on the Origin and Spread of Nationalism*. New York: Verso, 2016, 6.

43. Spiegel, James S. "The Ethics of Virtual Reality Technology: Social Hazards and Public Policy Recommendations." *Science and Engineering Ethics* 24, no. 5 (October 2018): 1537–1550. https://doi.org/10.1007/s11948-017-9979-y.

44. Reeves, Byron, and Clifford Ivar Nass. *The Media Equation: How People Treat Computers, Television, and New Media like Real People and Places*. Stanford: CSLI Publications, 1996.

45. Koburger, Daniel B. "Legal Implications of Public Spaces in Virtual Reality," September 22, 2019. https://www.americanbar.org/groups/intellectual_property_law/publications/landslide/2019-20/september-october/legal-implications-public-spaces-virtual-reality/.

46. Koburger, "Legal Implications of Public Spaces in Virtual Reality."

47. Bosse, Tibor, Tilo Hartmann, Romy A.M. Blankendaal, Nienke Dokter, Marco Otte, and Linford Goedschalk. "Virtually Bad: A Study on Virtual Agents That Physically Threaten Human Beings." In *Proceedings of the 17th International Conference on Autonomous Agents and MultiAgent Systems*, 1258–1266. AAMAS '18. Richland, SC: International Foundation for Autonomous Agents and Multiagent Systems, 2018.

48. Gutierrez, Carlos. "LGBTQ+ Virtual Reality: Equity and Inclusion Challenges in VR." *ARPost*, January 13, 2022. https://arpost.co/2022/01/13/lgbtq-virtual-reality-challenges-vr/.

49. Rubio-Licht, Nat. "How Virtual Reality Is Changing Sexual Harassment Training." *Protocol*, March 30, 2022. https://www.protocol.com/workplace/vr-sexual-harassment-training.

50. Verma, Pranshu. "Your next Sexual Harassment Training Could Be in Virtual Reality." *Washington Post*, April 19, 2022. https://www.washingtonpost.com/technology/2022/04/19/virtual-reality-sexual-harassment-training/.

51. O'Hagan, Joseph, Florian Mathis, and Mark McGill. "User Reviews as a Reporting Mechanism for Emergent Issues Within Social VR Communities." In *Proceedings of the 22nd International Conference on Mobile and Ubiquitous Multimedia*, 236–243. Vienna Austria: ACM, 2023. https://doi.org/10.1145/3626705.3627780.

Chapter 8

What Should VR Be?

When imagining the possible futures of new technologies, futurists often go to extremes. Either a technology will usher in a new golden age for humankind or it will destroy it. Imagining utopian and dystopian futures does serve a purpose, though: conceiving the beneficial applications of a technology helps to set goals for development and investment, while describing what can go horribly wrong prepares groups to mitigate against these possibilities long before they manifest into an uncontrollable problem.

The beginning of this book imagined a dystopian future, one in which people do not have an effective choice about whether or not to use VR, nor do they have control over what data about them is gathered and shared by the devices they use. That data would be used to manipulate their behaviors and decisions, and users would not be safe from those seeking to exploit them for financial gain or to inflict psychological harm. Economies of scale and the network effect lead to the concentration of resources into the hands of a few, while the value of most consumers is measured by the amount of time they spend in virtual worlds.

A possible utopian future is one in which these devices bring transformative experiences to more people around the world. Engineers are able to experiment with new designs quickly and cheaply, leading to innovations that benefit society. Medical professionals can plan surgeries in virtual spaces using 3D medical imaging, potentially saving lives. Students have access to visualizations of historical sites and events, and they each can have a front-row seat to world-class lectures. New forms of interactive entertainment place audiences in the middle of the action, encouraging people to imagine themselves as active participants in stories rather than passive observers. Meeting in virtual environments reduces the need for physical travel and the carbon footprint associated with it. New forms of social bonds emerge as people can explore other cultures and meet other people who can bring new perspectives to worldviews. Power and control over the technology underpinning VR are distributed, and people have alternative means of participation in a global culture even if they choose to refrain from using the technology.

While utopian futures are often funded by public relations campaigns and advertisements to attract the attention of investors and early adopters of new technologies, dystopian predictions are usually presented by academics, nonprofit organizations, and policymakers, who are more often than not met with a degree of public skepticism. Educator and social critic Neil Postman, however, cautioned that the adoption of any new technology is a Faustian bargain: "For every advantage a new technology offers, there is always a corresponding disadvantage."[1] Postman goes on to write in *The End of Education* that these advantages and disadvantages are not evenly dispersed or experienced: "This means that every new technology benefits some and harms others."[2]

An alternative construction of utopian and dystopian predictions of technologies is represented in Gartner's Hype Cycle. The multinational research company has popularized the notion that most new technologies begin with a "peak of inflated expectations," which is followed by a "trough of disillusionment" before they finally settle into a "plateau of productivity."[3] Gartner explains that this pattern is shaped by the initial publicity and excitement about a new technology and the possibilities that it can offer. As that excitement dies off or is revealed to be overenthusiastic, an overcorrection into disillusionment over the broken utopian promises gives way to further advancements of engineering that improve the technology. Finally, business adoption levels off to produce a longer and sustained practical use, after which businesses and individuals have begun to understand what uses the technology is good for.[4]

DEVELOPMENT HYPE CYCLES

The Gartner Hype Cycle is used by investors and businesses to adjust the timing of investment in a startup or the integration of a technology into their operations. Postman's prediction about technologies, though, focuses more on the people affected by a new technology more than businesses attempting to profit from it. Understanding the benefits and risks of adopting a new technology from the outset can prepare users for the possible shock that those risks create with that adoption. Like a Trojan Horse hiding the negative effects of a technology while its benefits are presented as a gift, the Faustian bargain warns people during the peak of inflated expectations to look for and expect risks so that they can set up mechanisms, limitations, and rules of use to limit those risks, lest they be taken by surprise.

VR brings with it multiple advantages over older technologies that many people can benefit from, but it also introduces new risks that, without external guidance and awareness by its users, can produce negative effects that have the potential to harm individuals and entire groups of people. Deciding how to implement VR, whether as a gaming portal, workplace tool, productivity aid, educational resource, communication device, therapeutic regimen, entertainment platform, or more likely, some combination of these uses and others, relies on a continuum of what is possible and what people allow. Using a free-market approach encourages developers to build and test a range of uses if they offer a promise of profit, with the most widely adopted options surviving the marketplace. While this may seem like a fair, democratic approach (consumers vote with their dollars!), often left unsaid is that many different marketplaces exist. Business-to-business is a different market than direct-to-consumer, healthcare industry, public education, and military contractors. Even within the direct-to-consumer space, that market is often differentiated by different demographic and psychographic groups. Developers may focus on one group while completely shutting out others, even if those other groups would benefit from its use. Despite its promise to the contrary, the marketplace is not rational.

A directed approach, conversely, constricts uses to those determined to be needed by a central authority for a particular purpose, whether as an employee control tool, healthcare insurance mitigator, or mass propaganda device. No matter whether there is demand for implementation, an authority can fund and distribute a VR application for purposes other than profit from the hardware or applications themselves. Though far-fetched, one could conceive of an authoritarian government or employer requiring citizens or employees to log into an HMD for a daily message or program. The authority can then use the headset's eye-tracking feature to confirm whether a person is viewing the program.[5] More likely, advertisers may subsidize a headset in exchange for a commitment to view ads.

Current marketing of VR devices already demonstrates that backers fall somewhere in the middle of this continuum. Meta's purchase of the Oculus Rift and rebranding it as the Meta Quest reflects Mark Zuckerberg's vision of using VR to extend the reach of his free, ad-supported social media and communication empire. Users can purchase games to play in the HMD or rent movies to watch in a virtual living room, but Meta's Horizon Worlds is free to use and is promoted as the future of social VR. The business model of this approach includes the branding of virtual spaces and the inclusion of paid messages, promising a future of VR in which content is provided to users largely for free but is determined by the advertisers and influencers paying for access.

Apple's Vision Pro, in contrast, markets its device as a personal productivity and entertainment tool. Combining the features of VR, AR, and mixed reality, Apple prefers to brand the experience as spatial computing, or the ability to complete tasks normally performed on an iPad or laptop in a fully immersive floating desktop. The high price of the headset positions it as a luxury good that supplements Apple's digital family of products, with the ability to seamlessly move between, for example, a MacBook Air, iPhone, iWatch, and Vision Pro. Apple is rumored to be planning the release of a lower-cost HMD in an effort to steal market share from Meta.[6] While users can access different apps and perform a variety of work in the Vision Pro, it is a closed ecosystem, often requiring an Apple iCloud account to sync workflow between its devices. In addition, unless pending government regulations and lawsuits change past practices, apps run on this device must be purchased through Apple's App Store, reinforcing the closed nature of software with Apple's products.

THE BIAS OF VR

These examples demonstrate the influence of developers, but VR as a medium exerts its own technological bias that both constrains and favors types of uses for VR and the manner of its use. As an immersive technology, VR can create an illusion of verisimilitude by providing multiple senses, including the visual, spatial audio, vestibular, and proprioception, with complementary stimuli that can simulate the experience of the real world. Combined with real-time, low-latency feedback, the awareness of the apparatus itself falls away as users of VR spend more time in the artificial environment. However, small discrepancies between sensing the real world and VR may lead to medium-specific effects, like differences in the perception of materials, objects, and people, judgments of distance and movement, and field of view discrepancies. These differences may create an uncanny feeling, or a psychological concern that an environment is almost real but not quite. Some users will be affected to a greater extent by motion sickness as a result of discrepancies between the visual and vestibular senses, preventing them from spending as much time, if any time at all, in virtual environments compared with others who are not affected. As long as VR remains a niche experience primarily for leisure, these differences have little consequence. However, as the technology becomes more integrated into everyday life, the expectation to use VR will increase, and the inability to tolerate VR perception may lead to a competitive disadvantage or even the loss of rights and access to resources.

One other bias of VR concerns the habituation of perception within the medium. As users spend more time in immersive environments, their perception of the real world may begin to feel more unreal. Disassociation of real-world spaces, people, and interactions may result as VR spaces begin to feel more normal to active users. A video posted to YouTube illustrates this phenomenon with a toddler playing with an iPad and a magazine. The toddler sits in front of an iPad and intuitively scrolls, zooms, and pages through words and images displayed on the device. The same toddler then sits in front of several magazines and becomes frustrated when the same actions have no effect on their pages. The toddler taps, slides, and pinches the pages, eventually pressing her finger against her own leg to confirm that her finger still works.[7] Likewise, heavy users of VR often report seeing an afterimage of the home screen after taking off their headset and feeling disoriented after being able to teleport to multiple spaces, making the real world seem less mobile by comparison. One reporter who spent two weeks using VR described brain/hand dissociation and balance issues, even waking up in the middle of the night and looking at his hands to confirm they were still there.[8]

Another common principle of new technologies is that they hardly ever are introduced as isolated entities. Instead, many precursor and concurrent innovations often create the foundation for a new technology to emerge and set up an environment for its flourishing. The current collection of VR devices could not exist without long-lasting, lightweight batteries, low-latency and high-resolution video displays, and accelerometers that can sense movement. In addition, real-time communication over the Internet, social media platforms, and game engine software help shape the information environment that builds the virtual spaces displayed in VR. The recent advancement of predictive and generative AI tools is making possible new features in VR and is helping to accelerate the design of new environments and NPCs that more closely mimic human behavior and communication. Examined separately, individual technologies often reveal only incremental changes to human experience. What can a VR game do beyond entertain the player interacting with it? However, each of these additional technologies carries with them their own biases, which could magnify issues present in VR if left unchecked.

When VR becomes more widely adopted, it begins to benefit from the network effect. The more users on a platform, the exponentially more valuable that platform becomes. Developers build spaces for platforms that have more people engaged on them, and people are more likely to join a platform where they can find more content and people over another that appears to be a virtual ghost town. Eventually, a platform becomes so valuable that the cost of opting out becomes too high. Citizens have a duty to participate in the governance over public spaces and resources, but consumers merely are invited (or not) to participate in promoting brands that lead to the purchase of goods and services. When the network effect creates a public square on a private platform, the question of who has the right to control that virtual space becomes paramount.

REGULATING THE FUTURE OF VR

The previous chapters of this book explore different areas in which VR can help or hurt human psychology and relationships, social organization, economic prosperity, and democratic institutions. Establishing ethical frameworks to engage this still relatively new technology will allow for development that considers the well-being of users, software developers, and platform producers. As this book has demonstrated, no one ethical theory adequately serves as the only system to organize a comprehensive policy for VR. Utilitarian ethics may help identify what would do the most good in some cases, while Kantian categorical imperatives can guide policies regarding transparency and bias. Virtue ethics may address issues like VR addiction and engagement, while an ethics of care can direct policies to prevent the harassment of members of marginalized groups. The rights of users may place limits on how their personal data is gathered and sold, and the rights of producers may guide how their devices and experiences are used. While designers have the creative freedom to choose what they produce and marketers are free to choose who should be the recipients of their incentives, an ethical code should prevent excluding classes of people from representation within experiences and unfair exclusion from the marketplace. Applying ethical theory to VR provides a systematic way of thinking through potential conflicts, evaluating the competing needs of stakeholders, and allowing for a consistent application of rules that prevents the most powerful or popular party from domineering a decision. By working through issues guided by ethical theory before a conflict arises, a veil of ignorance can be draped over decision-makers to engender trust with all parties affected by a decision.[9]

Sometimes, though, ethical theories conflict with one another when recommending the best approach for action. A categorical imperative to protect the privacy of users may clash against the need to reveal personal information to prevent abusers from inflicting harm on others. A utilitarian approach to benefit the most people by sharing VR data that may discover a cure for a disease may come into conflict with human rights ethics that protect the integrity of an individual's personal and sensitive information about their health. An ethic that moderates and automatically removes offensive content to protect children from experiencing it may conflict with the freedom for consenting adults to engage in any conduct they choose. Different cultures may have conflicting social norms for interactions. If an AI-guided translation agent has the power to change a message and behavior of an avatar in a social VR space, it could conform the social intent of the message to the culture of the person receiving it. However, that may also subvert the conscious will of people to control their own communication and prevent people from learning the nuances of different cultures, a valuable method of broadening the perspective of one's own understanding.

Even if it does no other good, an ethics of VR can provide a thermostatic regulator on the effects of the technology. By establishing a set of guiding principles that actively monitor and adjust the balance between technological advancement and its effects on people and institutions, a code of ethics advances the notion that the development and use of VR are aligned with human values. Such a code can act as a safeguard against potential abuses and negative consequences, promoting an integration of VR that enhances the human experience without sacrificing individual rights or overly impeding its development.

One key element of any ethical code, in order for it to be effective, is transparency. Opaque codes contained within internal review boards of companies may satisfy shareholders and corporate boards, but they do not quell the concerns of those affected by their decisions. Like the black box problem of AI, if the process by which decisions are made is not visible to those who are affected, there is less chance of accepting a decision as part of a social compact. A code of ethics ideally is developed by a community instead of the producer of a platform. To further shield the adjudication of ethical rules from undue influence, any organization designed to enforce a code should be independent. The removal of OpenAI board members who oversaw the ethical implementation of ChatGPT as a condition of the return of CEO Sam Altman[10] and Meta's dissolution of the responsible innovation team that kept guard against potential harms arising from the company's products[11] demonstrate the importance of independent boards that oversee the implementation of ethical conduct. The motivations, risks, and sacrifices of stakeholders, as well as the benefits a party stands to gain from a decision, all must be considered and balanced by those who oversee and promote the establishment of ethics that all parties follow.

While it is relatively easy to imagine extreme outcomes when thinking about possible futures resulting from the widespread use of VR, it is much more frustrating to realize that all new technologies bring with them unintended consequences. No matter the amount of deliberation that goes into establishing a code of ethics, not every possible scenario can be imagined. When Arthur C. Clarke envisioned in 1945 the possibility of space stations in geostationary orbit transmitting global communications, he described massive platforms with astronauts performing constant maintenance.[12] Despite his prescience, Clarke could not foresee the development of the transistor and microchip, obsolescing the need for vacuum tubes and allowing for the miniaturization of electronic devices, including satellites. There likely will be one or more uses for VR that no one has yet seen. As the technology miniaturizes from bulky headsets to glasses to contacts and possibly brain implants, developments in companion technologies may provide advances that will bring with them new challenges and risks.

Given this uncertainty, the goal of this book has not been to establish a finalized code of ethics for VR. Instead, posing questions about this immersive medium hopefully gives readers some practice in asking questions and considering different possibilities for the ethical uses and limits of VR. This is so that, when unanticipated consequences emerge, the decisions of what to do about the new risks and challenges do not fall solely to a small group of people. Only when all stakeholders who are free from pressure by those seeking undue influence are involved in these decisions will guidelines emerge that will be widely accepted.

While outright rejecting or banning VR is not a practical solution, it also is not helpful to assume that the adoption and development of VR will follow an inevitable path that cannot be modified, questioned, or critically assessed. The impacts of VR and other immersive technologies will be transformative, just as the Internet and social media have remade not only digital cultures but also older and more established institutions and human relationships themselves. Developments toward a dystopian future are far from inevitable. Ethical VR policy can serve as a toolset for users and developers of VR to employ and reimagine the medium as one in service of human psychological and social needs instead of humans becoming servants of it.

NOTES

1. Postman, Neil. *The End of Education: Redefining the Value of School*. New York: Vintage Books, 1996, 192.
2. Postman, *The End of Education*, 192.
3. Gartner Research. "Understanding Gartner's Hype Cycles." *Gartner*, August 20, 2018. https://www.gartner.com/en/documents/3887767.
4. Gartner Research, "Understanding Gartner's Hype Cycles."
5. Actually, it may not be so farfetched: Amazon already uses scanners, cameras, and AI systems to track its warehouse employees and penalize them for too much time off-task, as reported by Greene, Jay. "Amazon's Employee Surveillance Fuels Unionization Efforts: 'It's Not Prison, It's Work.'" *Washington Post*, December 14, 2021. https://www.washingtonpost.com/technology/2021/12/02/amazon-workplace-monitoring-unions/.
6. Lee, Chia-Han, and Jack Wu. "Vision Pro Gen 2 Rumored for 2026; Apple Searching for Less Expensive Micro OLEDs." *DIGITIMES*, May 8, 2024. https://www.digitimes.com/news/a20240507PD210/apple-apple-vision-pro-2026-micro-oled-panel.html.
7. "A Magazine Is an iPad That Does Not Work." *YouTube*. October 6, 2011. 1:25. https://www.youtube.com/watch?v=aXV-yaFmQNk.

8. Dedezade, Esat. "I Spent Two Weeks Working in VR and Now I'm Not Sure What's Real." *Wired UK*, February 28, 2021. https://www.wired.co.uk/article/working-in-vr-headset.

9. Rawls, John. *Justice as Fairness: A Restatement.* Cambridge, MA: Harvard University Press, 2001.

10. Wiggers, Kyle. "A Timeline of Sam Altman's Firing from OpenAI—and the Fallout." *TechCrunch*, January 5, 2024. https://techcrunch.com/2024/01/05/a-timeline-of-sam-altmans-firing-from-openai-and-the-fallout/.

11. Bell, Karissa. "Meta Dissolves Team Responsible for Discovering 'Potential Harms to Society' in Its Own Products." *Engadget*, September 8, 2022. https://www.engadget.com/meta-responsible-innovation-team-disbanded-194852979.html.

12. Tweney, Dylan. "May 25, 1945: Sci-Fi Author Predicts Future by Inventing It." *Wired*, May 25, 2011. https://www.wired.com/2011/05/0525arthur-c-clarke-proposes-geostationary-satellites/.

Bibliography

"About the Network." *Crash Override Network*. Accessed May 15, 2024. http://www.crashoverridenetwork.com/about.html.

Acxiom. "What Is Identity Resolution?" Accessed September 10, 2023. https://www.acxiom.com/identity-resolution/.

ADA.gov. "Guidance on Web Accessibility and the ADA," March 18, 2022. https://www.ada.gov/resources/web-guidance/.

Adhanom, Isayas Berhe, Paul MacNeilage, and Eelke Folmer. "Eye Tracking in Virtual Reality: A Broad Review of Applications and Challenges." *Virtual Reality* 27, no. 2 (June 2023): 1481–1505. https://doi.org/10.1007/s10055-022-00738-z.

Alan, Dershowitz. "When Is Doxxing Ethical or Unethical?" *The Hill* (blog), December 11, 2023. https://thehill.com/opinion/judiciary/4353133-when-is-doxxing-ethical-or-unethical/.

Altman, Sam. "Planning for AGI and Beyond." *OpenAI*, February 24, 2023. https://openai.com/blog/planning-for-agi-and-beyond.

Anderson, Benedict. *Imagined Communities: Reflections on the Origin and Spread of Nationalism*. New York: Verso, 2016.

Anderson, Janna, and Lee Rainie. "The Metaverse in 2040." *Pew Research Center*, June 30, 2022. https://www.pewresearch.org/internet/2022/06/30/the-metaverse-in-2040/.

Angwin, Julia, Jeff Larson, Surya Mattu, and Lauren Kirchner. "Machine Bias." *ProPublica*, May 23, 2016. https://www.propublica.org/article/machine-bias-risk-assessments-in-criminal-sentencing.

Anti-Defamation League Center for Technology & Society. "Online Hate and Harassment: The American Experience 2022." 2022. https://www.adl.org/sites/default/files/pdfs/2022-06/Online-Hate-and-Harassment-Survey-2022-v7.pdf.

Apple, Inc. "The App Store Turns 10." July 5, 2018. https://www.apple.com/newsroom/2018/07/app-store-turns-10/.

——. "Introducing Apple Vision Pro: Apple's First Spatial Computer." *Apple Newsroom*. Accessed April 12, 2024. https://www.apple.com/newsroom/2023/06/introducing-apple-vision-pro/.

Aristotle, and Anthony Kenny. *The Eudemian Ethics*. Oxford World's Classics. Oxford: Oxford University Press, 2011.

Aristotle, J.A.K. Thomson, and Hugh Tredennick. *The Nicomachean Ethics*. Penguin Classics. New York: Penguin Books, 2004.

ASMMC-MMAC'18: Proceedings of the Joint Workshop of the 4th Workshop on Affective Social Multimedia Computing and First Multi-Modal Affective Computing of Large-Scale Multimedia Data. New York: Association for Computing Machinery, 2018.

Atske, Sara. "Americans and Privacy: Concerned, Confused and Feeling Lack of Control Over Their Personal Information." *Pew Research Center: Internet, Science & Tech* (blog), November 15, 2019. https://www.pewresearch.org/internet/2019/11/15/americans-and-privacy-concerned-confused-and-feeling-lack-of-control-over-their-personal-information/.

Aulner, Francois, and Foo Yun Chee. "Google Loses Challenge against EU Antitrust Ruling, $2.8-Bln Fine." *Reuters*, November 10, 2021. https://www.reuters.com/technology/eu-court-upholds-eu-antitrust-ruling-against-google-2021-11-10/.

Bagchi, Saurabh. "Why We Need to See Inside AI's Black Box." *Scientific American*, May 26, 2023. https://www.scientificamerican.com/article/why-we-need-to-see-inside-ais-black-box/.

Baker, Harry. "PSVR 2 Sold 600k Units Up to April, Outpacing Original PSVR." *UploadVR*, May 24, 2023. https://www.uploadvr.com/psvr-2-600k-units-sold/.

Baker-White, Emily. "'This Is Blackface': Inside the Virtual Reality Company Trying to Scale Diversity Training." *BuzzFeed News*, December 1, 2021. https://www.buzzfeednews.com/article/emilybakerwhite/diversity-training-mursion-vr-white-actors.

Barlow, Rich. "Does Twitter, under Elon Musk, Need Government Regulation?" *BU Today*, November 18, 2022. https://www.bu.edu/articles/2022/do-elon-musk-and-twitter-need-government-regulation/.

Barr, Kyle. "Meta Quest Horizon OS Will Finally Make 3rd Party VR Headsets Worth Buying." *Gizmodo*, April 22, 2024. https://gizmodo.com/meta-quest-horizon-vr-quest-3xbox-1851426974.

Barry, Peter Brian. "#MeToo and the Ethics of Doxing Sexual Transgressors." In *The Palgrave Handbook of Sexual Ethics*, edited by David Boonin, 507–523. Cham: Springer International Publishing, 2022. https://doi.org/10.1007/978-3-030-87786-6_30.

Bayliss, Ben. "Deaf Players Create VRChat Community and Teach Sign Language." *DualShockers*, November 18, 2019. https://www.dualshockers.com/vrchat-deaf-community-teach-sign-language/.

BeAnotherLab. "The Machine to Be Another," 2022. https://beanotherlab.org/home/work/tmtba/.

Beilinson, Jerry. "Palmer Luckey and the Virtual Reality Resurrection." *Popular Mechanics*, May 28, 2014. https://www.popularmechanics.com/technology/gadgets/a12956/palmer-luckey-and-the-virtual-reality-resurrection-16834760/.

Belamire, Jordan. "My First Virtual Reality Groping." *Mic*, October 21, 2016. https://www.mic.com/articles/157415/my-first-virtual-reality-groping-sexual-assault-in-vr-harassment-in-tech-jordan-belamire.

Bell, Karissa. "Meta Dissolves Team Responsible for Discovering 'Potential Harms to Society' in Its Own Products." *Engadget*, September 8, 2022. https://www.engadget.com/meta-responsible-innovation-team-disbanded-194852979.html.

Bellinger, Daniel, Kristin Wehrmann, Anna Rohde, Maria Schuppert, Stefan Störk, Michael Flohr-Jost, Dominik Gall, et al. "The Application of Virtual Reality Exposure versus Relaxation Training in Music Performance Anxiety: A Randomized Controlled Study." *BMC Psychiatry* 23, no. 1 (August 1, 2023): 555. https://doi.org/10.1186/s12888-023-05040-z.

Benjamin, Ruha. *Race after Technology: Abolitionist Tools for the New Jim Code*. Medford, MA: Polity, 2019.

Bentham, Jeremy, Stephen G. Engelmann, and Philip Schofield. *Selected Writings*. Rethinking the Western Tradition. New Haven: Yale University Press, 2011.

Bloom, Nicholas, Jose Maria Barrero, Steven Davis, Brent Meyer, and Emil Mihaylov. "Survey: Remote Work Isn't Going Away—and Executives Know It." *Harvard Business Review*, August 28, 2023. https://hbr.org/2023/08/survey-remote-work-isnt-going-away-and-executives-know-it.

Bodyswaps. "Bodyswaps: Transforming Behaviours—for Good." Accessed September 21, 2023. https://bodyswaps.co/wp-content/uploads/2021/03/Bodyswaps-Learning-Framework.pdf.

Bonilla-Silva, Eduardo. *Racism without Racists: Color-Blind Racism and the Persistence of Racial Inequality in America*. 6th ed. Lanham, MD: Rowman & Littlefield, 2022.

Bosse, Tibor, Tilo Hartmann, Romy A.M. Blankendaal, Nienke Dokter, Marco Otte, and Linford Goedschalk. "Virtually Bad: A Study on Virtual Agents That Physically Threaten Human Beings." In *Proceedings of the 17th International Conference on Autonomous Agents and MultiAgent Systems*, 1258–1266. AAMAS'18. Richland, SC: International Foundation for Autonomous Agents and Multiagent Systems, 2018.

Boudet, Julien, Brian Gregg, Kathryn Rathje, Eli Stein, and Kai Vollhardt. "The Future of Personalization—and How to Get Ready for It." *McKinsey*, June 18, 2019. https://www.mckinsey.com/capabilities/growth-marketing-and-sales/our-insights/the-future-of-personalization-and-how-to-get-ready-for-it.

Bowman, Emma. "After Data Breach Exposes 530 Million, Facebook Says It Will Not Notify Users." *NPR*, April 9, 2021. https://www.npr.org/2021/04/09/986005820/after-data-breach-exposes-530-million-facebook-says-it-will-not-notify-users.

Buck, Stephanie. "Remembering the 90s Flame Wars: A Simpler Time of Cyberbullying." *Medium*, August 23, 2016. https://timeline.com/flame-wars-early-cyberbullying-1c509aa5ffad.

Bunz, Mercedes. "Are Hewlett-Packard Computers Really Racist?" *The Guardian*, December 23, 2009. https://www.theguardian.com/media/pda/2009/dec/23/hewlett-packard.

Burgess, Matt. "The Quiet Way Advertisers Are Tracking Your Browsing." *Wired*, February 26, 2022. https://www.wired.com/story/browser-fingerprinting-tracking-explained/.

Butler, Judith. *The Force of Nonviolence: An Ethico-Political Bind*. New York: Verso, 2020.

——. *Gender Trouble: Feminism and the Subversion of Identity*. New York: Routledge, 2006.

——. *Who's Afraid of Gender?* 1st ed. New York: Farrar, Straus and Giroux, 2024.

Cadwalladr, Carole, and Emma Graham-Harrison. "Revealed: 50 Million Facebook Profiles Harvested for Cambridge Analytica in Major Data Breach." *The Guardian*, March 17, 2018. https://www.theguardian.com/news/2018/mar/17/cambridge-analytica-facebook-influence-us-election.

CAIS, "Statement on AI Risk." May 30, 2023. https://www.safe.ai/statement-on-ai-risk.

Carvajal, Andrea. "Notes on Blindness Review." *VR Voice* (blog), August 31, 2019. https://vrvoice.co/notes-on-blindness-review/.

Castro, Chiara. "India Approves New Privacy Law, but Is It Really a Win for Citizens?" *TechRadar*, August 24, 2023. https://www.techradar.com/computing/cyber-security/india-approves-new-privacy-law-but-is-it-really-a-win-for-citizens.

Center for Countering Digital Hate. "Facebook's Metaverse." *CCDH*, December 30, 2021. https://counterhate.com/research/facebooks-metaverse/.

Cerovac, Ivan. *John Stuart Mill and Epistemic Democracy*. Lanham, MD: Lexington Books, 2022.

Chin, Josh, and Gillian Wong. "China's New Tool for Social Control: A Credit Rating for Everything." *Wall Street Journal*, November 28, 2016. http://www.wsj.com/articles/chinas-new-tool-for-social-control-a-credit-rating-for-everything-1480351590.

Christians, Clifford G., Mark Fackler, Kathy Brittain Richardson, and Peggy J. Kreshel. *Media Ethics: Cases and Moral Reasoning*. 11th ed. New York: Routledge, Taylor & Francis Group, 2020, 4–10.

Clegg, Nick. "Ensuring an Open and Interoperable Metaverse." *Meta* (blog), May 18, 2022. https://about.fb.com/news/2022/05/ensuring-an-open-and-interoperable-metaverse/.

Cline, Ernest. *Ready Player One*. New York: Crown Publishers, 2011.

Cogburn, Courtney D., Jeremy Bailenson, Elise Ogle, Tobin Asher, and Teff Nichols. "1000 Cut Journey." In *ACM SIGGRAPH 2018 Virtual, Augmented, and Mixed Reality*, 1–1. Vancouver British Columbia Canada: ACM, 2018. https://doi.org/10.1145/3226552.3226575.

Confucius, Roger T. Ames, and Henry Rosemont. *The Analects of Confucius: A Philosophical Translation*. Classics of Ancient China. New York: Ballantine Books, 1999.

Congressional Research Service. "Section 230: An Overview," April 7, 2021. https://crsreports.congress.gov/product/pdf/R/R46751.

Contreras, Russell. "New Virtual Reality Technology Seeks to Teach Racial Empathy." *Axios*, May 15, 2023. https://www.axios.com/2023/05/15/new-vr-role-playing-insight-racism.

Cooper, Mark, and Gene Kimmelman. "The Digital Divide Confronts The Telecommunications Act of 1996: Economic Reality Versus Public Policy." *Consumer Reports*, September 14, 1999. https://advocacy.consumerreports.org/wp-content/uploads/2013/04/telecom1-0299.pdf.

Cox, Alex. "The History of Minecraft: The Best Selling PC Game Ever." *TechRadar*, September 4, 2020. https://www.techradar.com/news/the-history-of-minecraft.

Craig, Emory. "Think Oculus Rift and Privacy Is an Issue? Just Wait until Mixed Reality Arrives," April 11, 2018. https://www.digitalbodies.net/oculus-rift-and-privacy-issue-just-wait-until-mixed-reality-arrives/.

Cross, Jason. "Apple Vision Pro: What Is Spatial Computing Anyway?" *Macworld*, February 2, 2024. https://www.macworld.com/article/2215989/apple-vision-pro-what-is-spatial-computing.html.

Cullinane, Michael. "The Ethics of Doxing Nazis." *Center for Digital Ethics & Society*, October 12, 2017. https://digitalethics.org/essays/ethics-doxing-nazis.

Damasio, Antonio R. *Descartes' Error: Emotion, Reason, and the Human Brain.* New York: Putnam, 1994.

Dave, Paresh. "Nvidia Chip Shortages Leave AI Startups Scrambling for Computing Power." *Wired*, August 24, 2023. https://www.wired.com/story/nvidia-chip-shortages-leave-ai-startups-scrambling-for-computing-power/.

Davis, Jenny L., Apryl Williams, and Michael W. Yang. "Algorithmic Reparation." *Big Data & Society* 8, no. 2 (July 2021): 205395172110448. https://doi.org/10.1177/20539517211044808, 8.

Davis, Phil. "Virtual Sex Software Spawns Lawsuit." *NBC News*, August 10, 2007. https://www.nbcnews.com/id/wbna20214184.

Dean, Grace. "AOC Says Elon Musk's Changes to Twitter's Blue Check Feature Are 'Ripe for Targeted Harassment.'" *Business Insider*, November 10, 2022. https://www.businessinsider.com/elon-musk-aoc-twitter-blue-check-notification-harassment-verified-subscription-2022-11.

Dedezade, Esat. "I Spent Two Weeks Working in VR and Now I'm Not Sure What's Real." *Wired UK*, February 28, 2021. https://www.wired.co.uk/article/working-in-vr-headset.

De Marzo, Stefano. "Munich-Based Atopia Secures Six-Figure Pre-Seed to up the Cultural Industry Game with VR Technologies." *EU-Startups*, August 30, 2023. https://www.eu-startups.com/2023/08/munich-based-atopia-secures-six-figure-pre-seed-to-up-the-cultural-industry-game-with-vr-technologies/.

DEV Studio. *Kavya Pearlman Guest Lecture: Past, Present, and Future of XR—XR Safety Initiative*, February 13, 2021. https://www.youtube.com/watch?v=Omi64iUqTPE.

Dibbell, Julian. "A Rape in Cyberspace: How an Evil Clown, a Haitian Trickster Spirit, Two Wizards, and a Cast of Dozens Turned a Database into a Society." *The Village Voice*, December 23, 1993. https://www.villagevoice.com/a-rape-in-cyberspace/.

D'Olimpio, Laura. "Ethics Explainer: Ethics of Care." *The Ethics Centre* (blog), May 16, 2019. https://ethics.org.au/ethics-explainer-ethics-of-care/.

Edmonds, David. *Would You Kill the Fat Man? The Trolley Problem and What Your Answer Tells Us about Right and Wrong*. Princeton, NJ: Princeton University Press, 2015.

European Parliament. "Artificial Intelligence Act: MEPs Adopt Landmark Law." March 13, 2024. https://www.europarl.europa.eu/news/en/press-room/20240308IPR19015/artificial-intelligence-act-meps-adopt-landmark-law.

——. "EU AI Act: First Regulation on Artificial Intelligence." August 6, 2023. https://www.europarl.europa.eu/news/en/headlines/society/20230601STO93804/eu-ai-act-first-regulation-on-artificial-intelligence.

Federal Deposit Insurance Corporation. "2021 FDIC National Survey of Unbanked and Underbanked Households." July 24, 2023. https://www.fdic.gov/analysis/household-survey/index.html.

Federal Trade Commission. "Fortnite Video Game Maker Epic Games to Pay More Than Half a Billion Dollars over FTC Allegations of Privacy Violations and Unwanted Charges." *Federal Trade Commission*, December 19, 2022. https://www.ftc.gov/news-events/news/press-releases/2022/12/fortnite-video-game-maker-epic-games-pay-more-half-billion-dollars-over-ftc-allegations.

——. "FTC Imposes $5 Billion Penalty and Sweeping New Privacy Restrictions on Facebook." *Federal Trade Commission*, July 24, 2019. https://www.ftc.gov/news-events/news/press-releases/2019/07/ftc-imposes-5-billion-penalty-sweeping-new-privacy-restrictions-facebook.

Feiner, Lauren. "Meta Acquisition of Within Reportedly Approved by Court in Loss for FTC." *CNBC*, February 1, 2023. https://www.cnbc.com/2023/02/01/ftc-loses-attempt-to-block-meta-acquisition-of-within.html.

Felsberg, Danielle T., Jaclyn P. Maher, and Christopher K. Rhea. "The State of Behavior Change Techniques in Virtual Reality Rehabilitation of Neurologic Populations." *Frontiers in Psychology* 10 (May 8, 2019): 979. https://doi.org/10.3389/fpsyg.2019.00979.

Ferguson, Christopher J., and Brad Glasgow. "Who Are GamerGate? A Descriptive Study of Individuals Involved in the GamerGate Controversy." *Psychology of Popular Media* 10, no. 2 (April 2021): 243–247. https://doi.org/10.1037/ppm0000280.

Fowler, Geoffrey A. "The Spy in Your Wallet: Credit Cards Have a Privacy Problem." *Washington Post*, August 29, 2019. https://www.washingtonpost.com/technology/2019/08/26/spy-your-wallet-credit-cards-have-privacy-problem/.

Freeman, Daniel, Polly Haselton, Jason Freeman, Bernhard Spanlang, Sameer Kishore, Emily Albery, Megan Denne, Poppy Brown, Mel Slater, and Alecia Nickless. "Automated Psychological Therapy Using Immersive Virtual Reality for Treatment of Fear of Heights: A Single-Blind, Parallel-Group, Randomised Controlled Trial." *The Lancet Psychiatry* 5, no. 8 (August 2018): 625–632. https://doi.org/10.1016/S2215-0366(18)30226-8.

Freeman, Guo, Samaneh Zamanifard, Divine Maloney, and Dane Acena. "Disturbing the Peace: Experiencing and Mitigating Emerging Harassment in Social Virtual Reality." *Proceedings of the ACM on Human-Computer Interaction* 6, no. CSCW1 (March 30, 2022): 1–30. https://doi.org/10.1145/3512932.

Freeman, Nate. "A History of Violence: Jordan Wolfson on His Shocking Foray into VR at the Whitney Biennial." *ARTnews.Com* (blog), March 1, 2017. https://www.artnews.com/art-news/artists/a-history-of-violence-jordan-wolfson-on-his-shocking-foray-into-vr-at-the-whitney-biennial-7856/.

Fultner, Barbara, and Jürgen Habermas, eds. *Jürgen Habermas: Key Concepts*. Durham: Acumen, 2011.

Funk, McKenzie. "The Man Who Trapped Us in Databases." *The New York Times*, September 22, 2023. https://www.nytimes.com/2023/09/22/magazine/hank-asher-data.html.

FusedVR Team. "Implementing VR Scene Transitions." *Game Developer*, June 13, 2017. https://www.gamedeveloper.com/design/implementing-vr-scene-transitions.

Gajsek, Dejan. "Virtual Reality Therapy as Exposure Therapy Alternative: Early Studies, Best Examples, and Possible Future of VR Therapy for Mental Health." *3D Development Bootcamp & XR Courses*, June 1, 2022. https://circuitstream.com/blog/virtual-reality-therapy/.

Gallese, Vittorio, and Michele Guerra, eds. *The Empathic Screen: Cinema and Neuroscience*. Oxford: Oxford University Press, 2020.

Gartner Research. "Understanding Gartner's Hype Cycles." *Gartner*, August 20, 2018. https://www.gartner.com/en/documents/3887767.

Gilligan, Carol. *In a Different Voice: Psychological Theory and Women's Development*. Cambridge, MA: Harvard University Press, 2003.

Gilligan, Carol, and David A.J. Richards. *The Deepening Darkness: Patriarchy, Resistance, and Democracy's Future*. New York: Cambridge University Press, 2009.

Goitein, Elizabeth. "The Government Can't Seize Your Digital Data. Except by Buying It." *Washington Post*, April 27, 2021. https://www.washingtonpost.com/outlook/2021/04/26/constitution-digital-privacy-loopholes-purchases/.

Goodman, Lizzy. "'Hunger In L.A.' Immerses Viewers In An Interactive Journalism Experience (And A Food Line)." *Fast Company*, January 31, 2012. https://www.fastcompany.com/1679530/hunger-in-la-immerses-viewers-in-an-interactive-journalism-experience-and-a-food-line.

Google. "Set up Google Cardboard." Accessed August 1, 2023. https://support.google.com/cardboard/answer/6295046.

Gore, Al. "Inauguration of the First World Telecommunication Development Conference (WTDC-94)." *Speech, Buenos Aires*, March 21, 1994. http://search.itu.int/history/HistoryDigitalCollectionDocLibrary/4.144.57.en.104.pdf.

Grabowski, Michael. "Perception and Poetics of VR Documentaries." Edited by Christian Iseli. *zDok.17: Being There*, November 2017, 1–12. https://blog.zhdk.ch/zdok/files/2017/11/referat_grabowski.pdf.

Greene, Jay. "Amazon's Employee Surveillance Fuels Unionization Efforts: 'It's Not Prison, It's Work.'" *Washington Post*, December 14, 2021. https://www.washingtonpost.com/technology/2021/12/02/amazon-workplace-monitoring-unions/.

Grewal, Paul. "Suspending Cambridge Analytica and SCL Group from Facebook." *Meta* (blog), March 17, 2018. https://about.fb.com/news/2018/03/suspending-cambridge-analytica/.

Gurman, Mark, and Julia Love. "Apple Nears Deal With OpenAI to Put ChatGPT on iPhone." *Bloomberg.com*, May 10, 2024. https://www.bloomberg.com/news/articles/2024-05-11/apple-closes-in-on-deal-with-openai-to-put-chatgpt-on-iphone.

Gutierrez, Carlos. "LGBTQ+ Virtual Reality: Equity and Inclusion Challenges in VR." *ARPost*, January 13, 2022. https://arpost.co/2022/01/13/lgbtq-virtual-reality-challenges-vr/.

Habermas, Jürgen, and Jürgen Habermas. *Moral Consciousness and Communicative Action*. Studies in Contemporary German Social Thought. Cambridge, MA: MIT Press, 2001.

Hagen, Alexis. "VR without Vision: Immersive Sensory System for People Who Are Blind." *Microsoft Research* (blog), April 7, 2020. https://www.microsoft.com/en-us/research/blog/bringing-virtual-reality-to-people-who-are-blind-with-an-immersive-sensory-based-system/.

Hauben, Michael, and Ronda Hauben. *Netizens: On the History and Impact of Usenet and the Internet*. Los Alamitos, CA: IEEE Computer Society Press, 1997.

Hayton, Phil. "Oculus Quest 2 Sales Jump Ahead of Xbox Series X and Series S." *PCGamesN*, June 8, 2022. https://www.pcgamesn.com/oculus/quest-2-meta-sales-xbox-series-x-s.

Heater, Brian. "After Losing Sight of Its Initial Pitch, Mojo Vision Eyes Pivot with $22.4M Raise." *TechCrunch*, April 5, 2023. https://techcrunch.com/2023/04/05/after-losing-sight-of-its-initial-pitch-mojo-vision-eyes-pivot-with-22-4m-raise/.

Heath, Alex. "This Is Meta's AR / VR Hardware Roadmap for the next Four Years." *The Verge*, March 1, 2023. https://www.theverge.com/2023/2/28/23619730/meta-vr-oculus-ar-glasses-smartwatch-plans.

Heaven, Will Douglass. "Welcome to the New Surreal: How AI-Generated Video Is Changing Film." *MIT Technology Review*, June 1, 2023. https://www.technologyreview.com/2023/06/01/1073858/surreal-ai-generative-video-changing-film/.

Heinaman, Robert, ed. *Plato and Aristotle's Ethics*. Ashgate Keeling Series in Ancient Philosophy. Burlington, VT: Ashgate, 2003.

Hill, Kashmir. "This Is Life in the Metaverse." *The New York Times*, October 7, 2022. https://www.nytimes.com/2022/10/07/technology/metaverse-facebook-horizon-worlds.html.

Hill, Kashmir, and Ryan Mac. "'Thousands of Dollars for Something I Didn't Do.'" *The New York Times*, March 31, 2023. https://www.nytimes.com/2023/03/31/technology/facial-recognition-false-arrests.html.

Hobbes, Thomas. *Leviathan*. Penguin Classics. Harmondsworth, Meddlesex: Penguin Books, 2017.

Hsu, Jeremy. "Why 'Uncanny Valley' Human Look-Alikes Put Us on Edge." *Scientific American*, April 3, 2012. https://www.scientificamerican.com/article/why-uncanny-valley-human-look-alikes-put-us-on-edge/.

Hu, Margaret. "Cambridge Analytica's Black Box." *Big Data & Society* 7, no. 2 (July 2020): 205395172093809. https://doi.org/10.1177/2053951720938091.

ICANN. "Bylaws for Internet Corporation for Assigned Names and Numbers: A California Nonprofit Public-Benefit Corporation—ICANN." Accessed April 16, 2024. https://www.icann.org/resources/pages/governance/bylaws-en/#article1.

——. "ICANN History Project—ICANN." Accessed April 16, 2024. https://www.icann.org/history.

IDC: The Premier Global Market Intelligence Company. "IDC—AR & VR Headsets Market Share." Accessed April 17, 2024. https://www.idc.com/promo/arvr.

Intel. "What Is a GPU? Graphics Processing Units Defined." Accessed September 1, 2023. https://www.intel.com/content/www/us/en/products/docs/processors/what-is-a-gpu.html.

The International Association of Privacy Professionals. "Web Beacon." Accessed September 10, 2023. https://iapp.org/resources/article/web-beacon/.

International Telecommunication Union. "Facts and Figures 2022: Latest on Global Connectivity amid Economic Downturn." *ITU Hub*, November 30, 2022. https://www.itu.int/hub/2022/11/facts-and-figures-2022-global-connectivity-statistics/.

Irwin, Neil, and Courtenay Brown. "A Beginner's Guide to the Fed's Big Conference in Jackson Hole." *Axios*, August 25, 2022. https://www.axios.com/2022/08/25/jackson-hole-fed-guide.

Iyer, Ravi, Spassena Koleva, Jesse Graham, Peter Ditto, and Jonathan Haidt. "Understanding Libertarian Morality: The Psychological Dispositions of Self-Identified Libertarians." Edited by Liane Young. *PLoS ONE* 7, no. 8 (August 21, 2012): e42366. https://doi.org/10.1371/journal.pone.0042366.

Janes, Barry T. "History and Structure of Public Access Television." *Journal of Film and Video* 39, no. 3 (1987): 14–23.

Jeff, Orlowski, dir. *The Social Dilemma*. Exposure Labs, 2020. https://netflix.com/title/81254224.

Juvrud, Joshua, Gabriel Ansgariusson, Patrik Selleby, and Magnus Johansson. "Game or Watch: The Effect of Interactivity on Arousal and Engagement in Video Game Media." *IEEE Transactions on Games* 14, no. 2 (June 2022): 308–317. https://doi.org/10.1109/TG.2021.3073084.

Kaiser, Brittany. *Targeted: My inside Story of Cambridge Analytica and How Trump, Brexit and Facebook Broke Democracy*. London: HarperCollins, 2019, 327.

Kang, Cecilia. "Facebook Grilled by Senators Over Its Effect on Children." *The New York Times*, September 30, 2021. https://www.nytimes.com/2021/09/30/technology/facebook-senate-hearing.html.

Kaplan, Fred. "Why Biden Wants to Keep the Law That Allows NSA Mass Surveillance, and Republicans Want to Kill It." *Slate*, March 1, 2023. https://slate.com/news-and-politics/2023/03/why-biden-wants-to-keep-section-702-nsa-mass-surveillance.html.

Kaplan, Isaac. "Jordan Wolfson's VR Work, 'Real Violence,' Has the Art World Talking." *Artsy*, March 17, 2017. https://www.artsy.net/article/artsy-editorial-gut-wrenching-vr-work-art-talking-violence.

Kemmerer, David. "What Modulates the Mirror Neuron System during Action Observation?" *Progress in Neurobiology* 205 (October 2021): 102128. https://doi.org/10.1016/j.pneurobio.2021.102128.

Kenedy, Jennifer, and Jorden Rutledge. "The NO FAKES Act: With Proposed Bill, Congress Set to Protect Against Unauthorized Digital Replicas of Faces, Names and Voices." *JD Supra*, October 17, 2023. https://www.jdsupra.com/legalnews/the-no-fakes-act-with-proposed-bill-2285269/.

Kharpal, Arjun. "Nvidia Warns More Semiconductor Curbs Will End U.S. Chipmakers' Ability to Compete in China." *CNBC*, August 24, 2023. https://www.cnbc.com/2023/08/24/nvidia-says-ai-chip-export-curbs-to-china-will-hit-us-chipmakers.html.

Kickstarter. "Oculus Rift: Step into the Game." January 30, 2016. https://www.kickstarter.com/projects/1523379957/oculus-rift-step-into-the-game.

Kim, Halla. *Kant and the Foundations of Morality*. Lanham, MD: Lexington Books, 2015.

King, Barbara J. "Can Cute Virtual Reality Spiders Help Reduce Arachnophobia?" *NPR*, September 1, 2016. https://www.npr.org/sections/13.7/2016/09/01/491991386/can-cute-virtual-reality-spiders-help-reduce-arachnophobia.

King, Hope. "Oculus Founder Rails against Critics on Reddit." CNNMoney, April 26, 2016. https://money.cnn.com/2016/04/26/technology/oculus-palmer-luckey/index.html.

Kirkcaldy, Andrew. "How Many Oculus Rifts Have Been Sold?" *PC Guide*, August 19, 2022. https://www.pcguide.com/vr/faq/how-many-oculus-rifts-have-been-sold/.

Klein, Alyson. "VR Devices Collect 'Intimate' Data, Lack Privacy Protections. Should Schools Invest?" *Education Week*, November 18, 2022. https://www.edweek.org/technology/vr-devices-collect-intimate-data-lack-privacy-protections-should-schools-invest/2022/11.

Koburger, Daniel B. "Legal Implications of Public Spaces in Virtual Reality." September 22, 2019. https://www.americanbar.org/groups/intellectual_property_law/publications/landslide/2019-20/september-october/legal-implications-public-spaces-virtual-reality/.

Kosoff, Maya. "'The Final Medium Is Virtual Reality': Palmer Luckey, the Black Sheep of Silicon Valley, Opens Up About V.R., Zuckerberg, and Why He Supports Trump." *Vanity Fair*, October 10, 2018. https://www.vanityfair.com/news/2018/10/the-final-medium-is-virtual-reality-palmer-luckey-the-black-sheep-of-silicon-valley-opens-up-about-vr-zuckerberg-and-why-he-supports-trump.

Kuperczko, Diana, Peter Kenyeres, Gergely Darnai, Norbert Kovacs, and Jozsef Janszky. "Sudden Gamer Death: Non-Violent Death Cases Linked to Playing Video Games." *BMC Psychiatry* 22, no. 1 (December 23, 2022): 824. https://doi.org/10.1186/s12888-022-04373-5.

Kushner, David. "Beyond Second Life: Philip Rosedale's Gutsy Plan for a New Virtual-Reality Empire." *IEEE Spectrum*, January 30, 2017. https://spectrum.ieee.org/beyond-second-life-philip-rosedales-gutsy-plan-for-a-new-virtualreality-empire.

Lang, Ben. "Understanding the Difference Between 'Screen Door Effect,' 'Mura,' & 'Aliasing.'" *Road to VR* (blog), February 21, 2018. https://www.roadtovr.com/whats-the-difference-between-screen-door-effect-sde-mura-aliasing-vr-headset/.

Lanier, Jaron. *Dawn of the New Everything: Encounters with Reality and Virtual Reality*. New York: Picador/Henry Holt and Company, 2018.

——. "VPL Research." *Virtual Reality Society*, September 8, 2013. https://www.vrs.org.uk/virtual-reality-profiles/vpl-research.html.

Lapowsky, Issie. "Biden Will Sign an Executive Order to Move Government Services Online." *Protocol*, December 13, 2021. https://www.protocol.com/bulletins/biden-executiv-order-government-modernize.

Leach, William. *Land of Desire: Merchants, Power, and the Rise of a New American Culture*. New York: Vintage Books, 1994, 37.

LeDoux, Joseph E. *Synaptic Self: How Our Brains Become Who We Are*. New York: Penguin Books, 2003.

Lee, Chia-Han, and Jack Wu. "Vision Pro Gen 2 Rumored for 2026; Apple Searching for Less Expensive Micro OLEDs." *DIGITIMES*, May 8, 2024. https://www.digitimes.com/news/a20240507PD210/apple-apple-vision-pro-2026-micro-oled-panel.html.

Lee, Claire Seungeun. "Analyzing Zoombombing as a New Communication Tool of Cyberhate in the COVID-19 Era." *Online Information Review* 46, no. 1 (January 25, 2022): 147–163. https://doi.org/10.1108/OIR-05-2020-0203.

Lee, Joyce, and Heekyong Yang. "Nvidia Supplier SK Hynix Says HBM Chips Almost Sold out for 2025." *Reuters*, May 2, 2024. https://www.reuters.com/technology/nvidia-supplier-sk-hynix-says-hbm-chips-almost-sold-out-2025-2024-05-02/.

Lee, Nicol Turner, Rashawn Ray, Samantha Lai, and Brooke Tanner. "Ensuring Equitable Access to AR/VR in Higher Education." *Brookings*, September 6, 2022. https://www.brookings.edu/articles/ensuring-equitable-access-to-ar-vr-in-higher-education/.

Leswing, Kif. "Nvidia Stock Falls after U.S. Government Restricts Chip Sales to China." *CNBC*, August 31, 2022. https://www.cnbc.com/2022/08/31/nvidia-stock-falls-after-us-government-restricts-chip-sales-to-china.html.

Levitz, Eric. "The 'Twitter Files' Is What It Claims to Expose." *Intelligencer*, December 10, 2022. https://nymag.com/intelligencer/2022/12/twitter-files-explained-elon-musk-taibbi-weiss-hunter-biden-laptop.html.

Levy, Steven. *Facebook: The Inside Story*. New York: Blue Rider Press, 2020.

Lifesize. "Lifesize 2019 Impact of Video Conferencing Report: More Collaborative Workplace Cultures Have Led to Spike in Video Communication." August 27, 2019. https://www.lifesize.com/press/impact-video-conferencing-report/.

Lih, Andrew. *The Wikipedia Revolution: How a Bunch of Nobodies Created the World's Greatest Encyclopedia*. New York: Hyperion, 2009.

Lin, Chiuhsiang Joe, and Susmitha Canny. "Effects of Virtual Target Size, Position, and Parallax on Vergence-Accommodation Conflict as Estimated by Actual Gaze." *Scientific Reports* 12, no. 1 (November 22, 2022): 20100. https://doi.org/10.1038/s41598-022-24450-9.

Linden Lab. "About." Accessed July 29, 2023. https://lindenlab.com/about.

Locke, John. *Two Treatises of Government*. Cambridge Texts in the History of Political Thought. New York: Cambridge University Press, 1988.

Lohr, Steve. "A Nation Ponders Its Growing Digital Divide." *The New York Times*, October 21, 1996. https://www.nytimes.com/1996/10/21/business/a-nation-ponders-its-growing-digital-divide.html.

Lomas, Natasha, and Romain Dillet. "Terms and Conditions Are the Biggest Lie of Our Industry." *TechCrunch* (blog), August 21, 2015. https://techcrunch.com/2015/08/21/agree-to-disagree/.

Ludlow, Edward. "OpenAI Nears $1 Billion of Annual Sales as ChatGPT Takes Off." *Bloomberg.Com*, August 29, 2023. https://www.bloomberg.com/news/articles/2023-08-30/openai-nears-1-billion-of-annual-sales-as-chatgpt-takes-off.

Luu, Wilson, Barbara Zangerl, Michael Kalloniatis, and Juno Kim. "Effects of Stereopsis on Vection, Presence and Cybersickness in Head-Mounted Display (HMD) Virtual Reality." *Scientific Reports* 11, no. 1 (June 11, 2021): 12373. https://doi.org/10.1038/s41598-021-89751-x.

Ma, Wayne. "Apple Hopes Chinese Suppliers Can Solve Vision Pro's Display Challenges." *The Information*, August 29, 2023. https://www.theinformation.com/articles/apple-hopes-chinese-suppliers-can-solve-vision-pros-display-challenges.

Maass, Anne, Mara Cadinu, and Silvia Galdi. "Sexual Harassment: Motivations and Consequences." In *The SAGE Handbook of Gender and Psychology*, edited by Michelle K. Ryan and Nyla R. Branscombe, 341–358. 1 Oliver's Yard, 55 City Road London EC1Y 1SP: SAGE Publications, Ltd, 2013. https://doi.org/10.4135/9781446269930.n21.

Machado, Caio C.V., and Thaís Helena Aguiar. "Emerging Regulations on Content Moderation and Misinformation Policies of Online Media Platforms: Accommodating the Duty of Care into Intermediary Liability Models." *Business and Human Rights Journal* 8, no. 2 (June 2023): 244–251. https://doi.org/10.1017/bhj.2023.25.

Madary, Michael, and Thomas K. Metzinger. "Real Virtuality: A Code of Ethical Conduct. Recommendations for Good Scientific Practice and the Consumers of VR-Technology." *Frontiers in Robotics and AI* 3 (2016). https://www.frontiersin.org/articles/10.3389/frobt.2016.00003.

Madison, James. "Founders Online: The Federalist No. 51, [6 February 1788]." *University of Virginia Press*. Accessed July 14, 2022. http://founders.archives.gov/documents/Hamilton/01-04-02-0199.

"A Magazine Is an iPad That Does Not Work." *YouTube*. October 6, 2011. 1:25. https://www.youtube.com/watch?v=aXV-yaFmQNk.

Manes, Stephen, and Paul Andrews. *Gates: How Microsoft's Mogul Reinvented an Industry and Made Himself the Richest Man in America*. New York: Simon and Schuster, 1994.

Martey, Rosa Mikeal, Jennifer Stromer-Galley, Jaime Banks, Jingsi Wu, and Mia Consalvo. "The Strategic Female: Gender-Switching and Player Behavior in Online Games." *Information, Communication & Society* 17, no. 3 (March 16, 2014): 286–300. https://doi.org/10.1080/1369118X.2013.874493.

Martin, Kirsten. "Predatory Predictions and the Ethics of Predictive Analytics." *Journal of the Association for Information Science and Technology* 74, no. 5 (May 2023): 531–545. https://doi.org/10.1002/asi.24743, 536.

Martins, Luis Bravo, and Samantha G. Wolfe. *Metaversed*. Indianapolis: John Wiley and Sons, 2022.

Matthews, Paul. "The Future of VR and AR." IEEE Computer Society, May 1, 2018. https://www.computer.org/publications/tech-news/trends/the-future-of-vr-and-ar/.

Mauran, Cecily. "Apple Vision Pro: ChatGPT Is Now Available on the $3,500 Headset." *Mashable*, February 6, 2024. https://mashable.com/article/apple-vision-pro-chatgpt-visionos-app.

McKinsey & Company. "The State of AI in 2023: Generative AI's Breakout Year." *McKinsey*, August 1, 2023. https://www.mckinsey.com/capabilities/quantumblack/our-insights/the-state-of-ai-in-2023-generative-AIs-breakout-year.

McLean, Willajeanne F. "Who Are You Wearing? Avatars, Blackface and Commodification of the Other." *Idea* 61, no. 3 (n.d.): 455–506, 485.

McLuhan, Marshall. *Understanding Media: The Extensions of Man*. 1st ed. Cambridge, MA: MIT Press, 1994.

McPhate, Mike. "California Today: In Virtual Reality, Investigating the Trayvon Martin Case." *The New York Times*, February 24, 2017. https://www.nytimes.com/2017/02/24/us/california-today-trayvon-martin-case-virtual-reality.html.

Meta Platforms, Inc. "Horizon Worlds: Virtual Reality Worlds and Communities." https://www.meta.com/horizon-worlds/.

——. "Making VR More Social with Meta Horizon Home." *Meta* (blog), June 14, 2022. https://about.fb.com/news/2022/06/making-vr-more-social-with-meta-horizon-home/.

——. "Meta Privacy Policy—How Meta Collects and Uses User Data." *Meta Privacy Center*, December 27, 2023. https://www.facebook.com/privacy/policy/.

——. "Meta Quest Account Payment Options." Accessed August 18, 2023. https://www.meta.com/help/quest/articles/accounts/payment-settings/meta-quest-payment-methods/.

——. "Meta Quest Supported Countries and Regions." *Meta Store*. Accessed May 13, 2024. https://www.meta.com/help/orders-and-returns/articles/quest-supported-countries/.

——. "Oculus Quest 2 Store: VR Games, Apps, & More." *Oculus*. Accessed August 24, 2023. https://www.oculus.com/experiences/quest/.

——. "Personal Space in Meta Horizon Worlds." Accessed September 1, 2023. https://www.meta.com/en-gb/help/quest/articles/horizon/explore-horizon-worlds/personal-space-in-meta-horizon-worlds/.

——. "VR Accessibility Design: Controller Mapping, Input, and Feedback." *Oculus Developers*. Accessed September 3, 2023. https://developer.oculus.com/resources/design-accessible-vr-controls/.

Microsoft. "Accessibility Technology & Tools." *Microsoft Accessibility*. Accessed September 3, 2023. https://www.microsoft.com/en-us/accessibility.

Middleton, Joe. "US Restricts Exports of Nvidia AI Chips to Middle East." *The Guardian*, August 31, 2023. https://www.theguardian.com/technology/2023/aug/31/us-restricts-exports-of-nvidia-ai-chips-to-middle-east.

Mill, John Stuart. *The Basic Writings of John Stuart Mill*. 2002 Modern Library pbk. ed. New York: Modern Library, 2002.

Miller, Mark Roman, Fernanda Herrera, Hanseul Jun, James A. Landay, and Jeremy N. Bailenson. "Personal Identifiability of User Tracking Data during Observation of 360-Degree VR Video." *Scientific Reports* 10, no. 1 (October 15, 2020): 17404. https://doi.org/10.1038/s41598-020-74486-y.

Miloff, Alexander, Philip Lindner, Peter Dafgård, Stefan Deak, Maria Garke, William Hamilton, Julia Heinsoo, et al. "Automated Virtual Reality Exposure Therapy for Spider Phobia vs. in-Vivo One-Session Treatment: A Randomized Non-Inferiority Trial." *Behaviour Research and Therapy* 118 (July 2019): 130–140. https://doi.org/10.1016/j.brat.2019.04.004.

"Minecraft Server List." *MinecraftServers.org*. Accessed July 30, 2023. https://minecraftservers.org/.

Minow, Newton. "American Rhetoric: Newton Minow—Address to the National Association of Broadcasters (Television and the Public Interest)." Accessed September 24, 2023. https://www.americanrhetoric.com/speeches/newtonminow.htm.

Monteiro, A. Reis. *Ethics of Human Rights*. New York: Springer, 2014.

Mulenga, Natasha. "Adele on Cultural Appropriation: 'I Didn't Read the F**king Room.'" *Teen Vogue*, October 8, 2021. https://www.teenvogue.com/story/adele-addressed-cultural-appropriation-backlash-bantu-knots-vogue-covers.

Murphy, Hannah. "How Will Facebook Keep Its Metaverse Safe for Users?" *Financial Times*, November 12, 2021. https://www.ft.com/content/d72145b7-5e44-446a-819c-51d67c5471cf.

Murray, Scott. "How Artificial Intelligence Can Help Combat Systemic Racism." *MIT News*, March 16, 2022. https://news.mit.edu/2022/how-ai-can-help-combat-systemic-racism-0316.

Nakamura, Lisa. *Cybertypes: Race, Ethnicity, and Identity on the Internet*. New York: Routledge, 2002.

——. *Digitizing Race: Visual Cultures of the Internet*. Minneapolis: University of Minnesota Press, 2008.

——. "Feeling Good about Feeling Bad: Virtuous Virtual Reality and the Automation of Racial Empathy." *Journal of Visual Culture* 19, no. 1 (April 2020): 47–64. https://doi.org/10.1177/1470412920906259.

Nast, Condé. "Met Gala 2023: Celebrities, Red Carpet, Theme & More." *Vogue*. Accessed September 1, 2023. https://www.vogue.com/tag/event/met-gala.

NBC News. "1980s Flashback: When Cell Phones Were a Novelty." *NBCNews.com*, June 16, 2015. https://www.nbcnews.com/watch/flashback/1980s-flashback-when-cell-phones-were-a-novelty-465541187745.

"A New (Virtual) Reality: The Rise of VR Headsets & EEG." *NeuroSky*, April 28, 2016. https://neurosky.com/2016/04/a-new-virtual-reality-the-rise-of-vr-headsets-eeg/.

Newport, Frank. "75 Years Ago, the First Gallup Poll." *Gallup.com*, October 20, 2010. https://news.gallup.com/opinion/polling-matters/169682/years-ago-first-gallup-poll.aspx.

Noble, Safiya Umoja. *Algorithms of Oppression: How Search Engines Reinforce Racism*. New York: New York University Press, 2018.

Noble, Safiya Umoja, and Sarah T. Roberts. "Through Google-Colored Glass(es)." In *Emotions, Technology, and Design*, 187–212. Elsevier, 2016. https://doi.org/10.1016/B978-0-12-801872-9.00009-0.

Norlock, Kathryn. "Feminist Ethics." In *The Stanford Encyclopedia of Philosophy*, edited by Edward N. Zalta, Summer 2019. Metaphysics Research Lab, Stanford University, 2019. https://plato.stanford.edu/archives/sum2019/entries/feminism-ethics/.

Norris, Pippa. *Digital Divide: Civic Engagement, Information Poverty, and the Internet Worldwide*. Cambridge University Press, 2001. https://doi.org/10.1017/CBO9781139164887.

Novet, Jordan. "Mark Zuckerberg Envisions a Billion People in the Metaverse Spending Hundreds of Dollars Each." *CNBC*, June 22, 2022. https://www.cnbc.com/2022/06/22/mark-zuckerberg-envisions-1-billion-people-in-the-metaverse.html.

Nowak, Kristine L., and Jesse Fox. "Avatars and Computer-Mediated Communication: A Review of the Definitions, Uses, and Effects of Digital Representations on Communication." *Review of Communication Research* 6 (2018): 30–53. https://doi.org/10.12840/issn.2255-4165.2018.06.01.015.

"NYT VR: How to Experience a New Form of Storytelling from The Times." *The New York Times*, November 5, 2015. https://www.nytimes.com/2015/11/08/magazine/nyt-vr-how-to-experience-a-new-form-of-storytelling-from-the-times.html.

Office of the Assistant Secretary for Health. "Surgeon General Issues New Advisory About Effects Social Media Use Has on Youth Mental Health." *HHS.gov*, May 23, 2023. https://www.hhs.gov/about/news/2023/05/23/surgeon-general-issues-new-advisory-about-effects-social-media-use-has-youth-mental-health.html.

O'Hagan, Joseph, Florian Mathis, and Mark McGill. "User Reviews as a Reporting Mechanism for Emergent Issues Within Social VR Communities." In *Proceedings of the 22nd International Conference on Mobile and Ubiquitous Multimedia*, 236–243. Vienna Austria: ACM, 2023. https://doi.org/10.1145/3626705.3627780.

O'Neil, Cathy. *Weapons of Math Destruction: How Big Data Increases Inequality and Threatens Democracy*. New York: Crown, 2016.

OpenAI. "Introducing ChatGPT." *OpenAI*, November 30, 2022. https://openai.com/blog/chatgpt.

——. "The Power of Continuous Learning." *OpenAI*, December 23, 2022. https://openai.com/blog/the-power-of-continuous-learning.

Outlaw, Jessica. "Virtual Harassment: The Social Experience of 600+ Regular Virtual Reality (VR) Users." *The Extended Mind*, April 4, 2018. https://www.extendedmind.io/the-extended-mind-blog/2018/04/04/2018-4-4-virtual-harassment-the-social-experience-of-600-regular-virtual-reality-vrusers.

Ovide, Shira. "The Big Deal in Amazon's Antitrust Case." *The New York Times*, May 25, 2021. https://www.nytimes.com/2021/05/25/technology/amazon-antitrust-lawsuit.html.

Owen, Malcolm. "Epic vs. Apple: Fortnite, History, Court Ruling, Appeal Results." *AppleInsider*, April 24, 2023. https://appleinsider.com/articles/20/08/23/apple-versus-epic-games-fortnite-app-store-saga----the-story-so-far.

Palmer, Philip. "The Best Minecraft Servers." *PC Gamer*, November 1, 2022. https://www.pcgamer.com/the-best-minecraft-servers/.

Park, Sang-Min, and Young-Gab Kim. "A Metaverse: Taxonomy, Components, Applications, and Open Challenges." *IEEE Access* 10 (2022): 4209–4251. https://doi.org/10.1109/ACCESS.2021.3140175.

Parrish, Audrey E. "Visual Illusions: Insights from Comparative Cognition." In *Comparative Cognition*, edited by James R. Anderson and Hika Kuroshima, 15–30. Singapore: Springer Singapore, 2021. https://doi.org/10.1007/978-981-16-2028-7_2.

Patel, Nilay. "Google and YouTube Are Trying to Have It Both Ways with AI and Copyright." *The Verge*, August 22, 2023. https://www.theverge.com/2023/8/22/23841822/google-youtube-ai-copyright-umg-scraping-universal.

Paul, Cassaundra. "How To Play Fortnite On VR." *CellularNews* (blog), July 21, 2023. https://cellularnews.com/mobile-accessories/how-to-play-fortnite-on-vr/.

PayPal. "Opening a PayPal Account." *PayPal User Agreement*, May 17, 2023. https://www.paypal.com/legalhub/useragreement-full.

Pearce, Celia. *Communities of Play: Emergent Cultures in Multiplayer Games and Virtual Worlds*. Cambridge, MA: MIT Press, 2009.

Perera, Ayesh. "What Is The McGurk Effect?" *Simply Psychology*, September 7, 2023. https://www.simplypsychology.org/mcgurk-effect.html.

Pimentel, Daniel, Maxwell Foxman, Donna Z. Davis, and David M. Markowitz. "Virtually Real, But Not Quite There: Social and Economic Barriers to Meeting Virtual Reality's True Potential for Mental Health." *Frontiers in Virtual Reality* 2 (February 17, 2021): 627059. https://doi.org/10.3389/frvir.2021.627059.

Plato. "Phaedrus." Translated by Benjamin Jowett, 2008. https://www.gutenberg.org/files/1636/1636-h/1636-h.htm.

Postman, Neil. *Building a Bridge to the 18th Century: How the Past Can Improve Our Future*. New York: Alfred A. Knopf, 1999, 42–53.

——. *The End of Education: Redefining the Value of School*. New York: Vintage Books, 1996.

——. *Technopoly: The Surrender of Culture to Technology*. New York: Vintage Books, 1993.

Pyasik, Maria, Gaetano Tieri, and Lorenzo Pia. "Visual Appearance of the Virtual Hand Affects Embodiment in the Virtual Hand Illusion." *Scientific Reports* 10, no. 1 (March 25, 2020): 5412. https://doi.org/10.1038/s41598-020-62394-0.

Quinn, Zoë. *Crash Override: How Gamergate (Nearly) Destroyed My Life, and How We Can Win the Fight against Online Hate*. New York: PublicAffairs, 2017.

Ramachandran, V.S. *The Tell-Tale Brain: A Neuroscientist's Quest for What Makes Us Human*. New York: W.W. Norton, 2011.

Rawls, John. *Justice as Fairness: A Restatement*. Cambridge, MA: Harvard University Press, 2001.

——. *A Theory of Justice. Rev. Ed.* Cambridge, MA: Belknap Press, 2005.

Redohl, Sarah. "Vuze+ 3D 360 Camera: Review, Best Practices and How to Use It." *Immersive Shooter*, August 27, 2018. https://www.immersiveshooter.com/2018/08/27/vuze-3d-360-camera-review-how-to/.

Reeves, Byron, and Clifford Ivar Nass. *The Media Equation: How People Treat Computers, Television, and New Media like Real People and Places*. Stanford: CSLI Publications, 1996.

Reporters Without Borders. "The Uncensored Library." Accessed August 1, 2023. https://www.uncensoredlibrary.com/.

Reznik, Audrey, Troy Nelson, Kaitlyn Abdo, and Christina Xu. "Why GPUs Are Essential for AI and High-Performance Computing." *Red Hat Developer*, November 21, 2022. https://developers.redhat.com/articles/2022/11/21/why-gpus-are-essential-computing.

Rheingold, Harold. *Virtual Reality*. New York: Touchstone/Simon & Schuster, 1992.

Robehmed, Natalie. "Highest-Paid YouTube Stars 2018: Markiplier, Jake Paul, PewDiePie and More." *Forbes*, December 3, 2018. https://www.forbes.com/sites/natalierobehmed/2018/12/03/highest-paid-youtube-stars-2018-markiplier-jake-paul-pewdiepie-and-more/.

Rosenberg, Matthew, Nicholas Confessore, and Carole Cadwalladr. "How Trump Consultants Exploited the Facebook Data of Millions." *The New York Times*, March 17, 2018. https://www.nytimes.com/2018/03/17/us/politics/cambridge-analytica-trump-campaign.html.

Rosman, Katherine. "The Latest High School Prank? It's a Snooze." *The New York Times*, December 23, 2021, sec. Style. https://www.nytimes.com/2021/12/23/style/school-sleep-prank.html.

Roth, Lorna. "Looking at Shirley, the Ultimate Norm: Colour Balance, Image Technologies, and Cognitive Equity." *Canadian Journal of Communication* 34, no. 1 (March 30, 2009): 111–136. https://doi.org/10.22230/cjc.2009v34n1a2196.

Rubin, April. "A Minecraft Player Set Out to Build the Known Universe, Block by Block." *The New York Times*, October 9, 2022. https://www.nytimes.com/2022/10/09/technology/minecraft-universe-developer.html.

Rubio-Licht, Nat. "How Virtual Reality Is Changing Sexual Harassment Training." *Protocol*, March 30, 2022. https://www.protocol.com/workplace/vr-sexual-harassment-training.

——. "The Metaverse Is on Its Way, According to Mark Zuckerberg." *Protocol*, March 16, 2022. https://www.protocol.com/mark-zuckerberg-metaverse-sxsw.

Sakiyama, Tomoko, Aisato Sasaki, and Yukio-Pegio Gunji. "Origin of Kanizsa Triangle Illusion." In *Soft Computing in Machine Learning*, edited by Sang-Yong Rhee, Jooyoung Park, and Atsushi Inoue, 273: 95–103. Advances in Intelligent Systems and Computing. Cham: Springer International Publishing, 2014. https://doi.org/10.1007/978-3-319-05533-6_10.

Salmanowitz, Natalie. "The Impact of Virtual Reality on Implicit Racial Bias and Mock Legal Decisions." *Journal of Law and the Biosciences* 5, no. 1 (May 1, 2018): 174–203. https://doi.org/10.1093/jlb/lsy005.

Salter, Michael. "From Geek Masculinity to Gamergate: The Technological Rationality of Online Abuse." *Crime, Media, Culture: An International Journal* 14, no. 2 (August 2018): 247–264, 259. https://doi.org/10.1177/1741659017690893.

Santos, Marc Ericson. "Code of Ethics for Virtual Reality." *Inborn Experience (UX in AR/VR)* (blog), October 10, 2019. https://medium.com/inborn-experience/code-of-ethics-for-virtual-reality-ebbdc5ae8ebc.

Schaefer, Kendra. "China's Corporate Social Credit System." *Trivium China*, November 16, 2020. https://www.uscc.gov/sites/default/files/2020-12/Chinas_Corporate_Social_Credit_System.pdf.

Schapiro, Zachary S. "Update: Processing Sensitive Personal Information under U.S. State Privacy Laws." *The National Law Review*, September 12, 2023. https://www.natlawreview.com/article/update-processing-sensitive-personal-information-under-us-state-privacy-laws.

Schneier, Bruce. *Data and Goliath: The Hidden Battles to Collect Your Data and Control Your World*. New York: W.W. Norton & Company, 2015.

Schofield, Philip. *Utility and Democracy: The Political Thought of Jeremy Bentham*. Oxford: Oxford University Press, 2009.

Schwartz, Alexandra. "Confronting the 'Shocking' Virtual-Reality Artwork at the Whitney Biennial." *The New Yorker*, March 20, 2017. https://www.newyorker.com/culture/cultural-comment/confronting-the-shocking-virtual-reality-artwork-at-the-whitney-biennial.

Schwartz, Bennett L., and John H. Krantz. *Sensation & Perception*. Second edition. Los Angeles: SAGE, 2019.

Second Life Community. "Inworld Employment." July 29, 2022. https://community.secondlife.com/forums/forum/312-inworld-employment/.

——. "Linden Homes Security System." April 26, 2019. https://community.secondlife.com/forums/topic/436103-%E2%9A%A1-linden-homes-security-system-%E2%9A%A1/.

——. "Number of Second Life Users?" May 24, 2021. https://community.secondlife.com/forums/topic/472381-number-of-second-life-users/.

Shah, Agam. "Intel's Lack of Potent Graphics May Hurt in VR and AI Chase." *PC World*, August 18, 2016. https://www.pcworld.com/article/416087/lack-of-a-high-quality-gpu-may-hold-intel-back-in-its-ai-and-vr-chase.html.

Shell, Susan Meld, ed. *The Strauss-Kruger Correspondence: Returning to Plato through Kant*. New York: Springer Science and Business Media, 2018.

Shipps, Alex. "Simulating Discrimination in Virtual Reality." *MIT News*, January 5, 2023. https://news.mit.edu/2023/simulating-discrimination-virtual-reality-0105.

Shirky, Clay. *Here Comes Everybody: The Power of Organizing without Organizations*. New York: Penguin Press, 2008.

Sim, May. *Remastering Morals with Aristotle and Confucius*. New York: Cambridge University Press, 2007.

"SL Bar Association." Accessed January 8, 2024. http://slbarassn.ning.com/.

Smith, Craig S. "Hallucinations Could Blunt ChatGPT's Success." *IEEE Spectrum*, March 13, 2023. https://spectrum.ieee.org/ai-hallucination.

Spiegel, James S. "The Ethics of Virtual Reality Technology: Social Hazards and Public Policy Recommendations." *Science and Engineering Ethics* 24, no. 5 (October 2018): 1537–1550. https://doi.org/10.1007/s11948-017-9979-y.

Stallman, Richard, and Richard M. Stallman. *Free Software, Free Society: Selected Essays*. Edited by Joshua Gay. Boston, Mass: Free Software Foundation, 2002.

Standage, Tom. *The Victorian Internet: The Remarkable Story of the Telegraph and the Nineteenth Century's on-Line Pioneers*. New York: Bloomsbury, 2014.

Stauffert, Jan-Philipp, Florian Niebling, and Marc Erich Latoschik. "Latency and Cybersickness: Impact, Causes, and Measures. A Review." *Frontiers in Virtual Reality* 1 (November 26, 2020): 582204. https://doi.org/10.3389/frvir.2020.582204.

Stephenson, Neal. *Snow Crash*. Bantam Spectra Book. New York: Bantam Books, 2008.

Stoll, Clifford. *Silicon Snake Oil: Second Thoughts on the Information Highway*. New York: Doubleday, 1996.

Suomela, Todd, Florence Chee, Bettina Berendt, and Geoffrey Rockwell. "Applying an Ethics of Care to Internet Research: Gamergate and Digital Humanities." *Digital Studies/Le Champ Numérique* 9, no. 1 (February 13, 2019): 4. https://doi.org/10.16995/dscn.302.

Swaak, Taylor. "What Colleges' Contracts for 'Metaversities' Leave Out." *The Chronicle of Higher Education*, June 17, 2022. https://www.chronicle.com/article/will-virtual-reality-metaversities-risk-students-data-privacy-we-asked-for-contracts-to-find-out.

Tang, Lu, Kayo Fujimoto, Muhammad (Tuan) Amith, Rachel Cunningham, Rebecca A. Costantini, Felicia York, Grace Xiong, Julie A. Boom, and Cui Tao. "'Down the Rabbit Hole' of Vaccine Misinformation on YouTube: Network Exposure Study." *Journal of Medical Internet Research* 23, no. 1 (January 5, 2021): e23262. https://doi.org/10.2196/23262.

"A Telegraph Monopoly." *The New York Times*, May 27, 1884, 4.

Thai, Kien T. P., Sungchul Jung, and Robert W. Lindeman. "On the Use of 'Active Breaks' to Perform Eye Exercises for More Comfortable VR Experiences." In *2020 IEEE Conference on Virtual Reality and 3D User Interfaces Abstracts and Workshops (VRW)*, 468–476. Atlanta, GA, USA: IEEE, 2020. https://doi.org/10.1109/VRW50115.2020.00099.

Thorbecke, Catherine. "OpenAI Launches a Version of ChatGPT for Businesses." *CNN*, August 28, 2023. https://www.cnn.com/2023/08/28/tech/chatgpt-enterprise-openai/index.html.

Tilley, Elspeth. "New Culture/Old Ethics: What Technological Determinism Can Teach Us about New Media and Public Relations Ethics." In *The Ethics of Emerging Media: Information, Social Norms, and New Media Technology*, edited by Bruce E. Drushel and Kathleen M. German, 191–212. New York: Continuum, 2011.

TODAY.com. "'What Is Internet?' Katie Couric, Bryant Gumbel Are Puzzled." Accessed September 3, 2023. https://www.today.com/video/-what-is-internet-katie-couric-bryant-gumbel-are-puzzled-62308421624.

Totilo, Stephen. "VR Project Puts Users in a Different Body to Sense What Gender Dysphoria Might Feel Like." *Axios*, July 27, 2023. https://www.axios.com/2023/07/27/vr-body-of-mine-transgender-experience.

Tricker, Brandy. "Taming the Wild West: Solving Virtual World Disputes Using Non-Virtual Law." *Rutgers Computer and Technology Law Journal* 35, no. 1 (September 22, 2008). https://www.thefreelibrary.com/Taming+the+Wild+West%3A+solving+virtual+world+disputes+using+non-virtual...-a0207060728.

Turkle, Sherry. *Life on the Screen: Identity in the Age of the Internet*. New York: Simon & Schuster, 1995.

Tweney, Dylan. "May 25, 1945: Sci-Fi Author Predicts Future by Inventing It." *Wired*, May 25, 2011. https://www.wired.com/2011/05/0525arthur-c-clarke-proposes-geostationary-satellites/.

U.S. Congress. Telecommunications Act of 1996, Pub. L. No. 104, § 706, 104 (1996). https://www.congress.gov/104/plaws/publ104/PLAW-104publ104.pdf.

U.S. National Archives and Records Administration. "Declaration of Independence: A Transcription." *National Archives*, November 1, 2015. https://www.archives.gov/founding-docs/declaration-transcript.

Union Avatars. "Realistic Full Body Avatar Maker." Accessed August 6, 2023. https://unionavatars.com/.

United Nations General Assembly. "Universal Declaration of Human Rights." *United Nations*, 1948. https://www.un.org/en/about-us/universal-declaration-of-human-rights.

Vaidhyanathan, Siva. *Antisocial Media: How Facebook Disconnects US and Undermines Democracy*. New York: Oxford University Press, 2018.

Vallance, Chris. "Police Investigate Virtual Sex Assault on Girl's Avatar." *BBC News*, January 2, 2024. https://www.bbc.com/news/technology-67865327.

Van Den Ende, J., W. Ravesteijn, and D. De Wit. "Shaping the Early Development of Television." *IEEE Technology and Society Magazine* 16, no. 4 (1997): 13–26. https://doi.org/10.1109/44.642518.

Vassilakopoulou, Polyxeni, and Eli Hustad. "Bridging Digital Divides: A Literature Review and Research Agenda for Information Systems Research." *Information Systems Frontiers* 25, no. 3 (June 2023): 955–969. https://doi.org/10.1007/s10796-020-10096-3.

Vendetta, Twisted. "Your Second Life Could Be Changing Soon (Alternate Title: The Clothes Can Stay on the Rack for the Day)." *The Twisted Perspective* (blog), October 23, 2008. https://twistedandtawdry.wordpress.com/2008/10/23/your-second-life-could-be-changing-soon-alternate-title-the-clothes-can-stay-on-the-rack-for-the-day/.

Verma, Pranshu. "Your next Sexual Harassment Training Could Be in Virtual Reality." *Washington Post*, April 19, 2022. https://www.washingtonpost.com/technology/2022/04/19/virtual-reality-sexual-harassment-training/.

Villegas-Galaviz, Carolina, and Kirsten Martin. "Moral Distance, AI, and the Ethics of Care." *AI & Society*, March 23, 2023. https://doi.org/10.1007/s00146-023-01642-z.

Vinney, Cynthia. "Gilligan's Ethics of Care." *ThoughtCo*, June 30, 2019. https://www.thoughtco.com/ethics-of-care-4691476.

Vogels, Emily A. "The Longer and More Often People Use Facebook, the More Ad Preferences the Site Lists about Them." *Pew Research Center*, December 3, 2019. https://www.pewresearch.org/short-reads/2019/12/03/facebook-ad-preferences-linked-to-frequency-of-use-age-of-account/.

———. "The State of Online Harassment." *Pew Research Center: Internet, Science & Tech* (blog), January 13, 2021. https://www.pewresearch.org/internet/2021/01/13/the-state-of-online-harassment/.

Waude, Adam. "Ambient Influence: How Background Music In Stores And Supermarkets Influences Consumers' Purchasing Decisions," February 23, 2016. https://www.psychologistworld.com/behavior/ambient-music-retail-psychological-arousal-customers.

Weinshel, Ben, Miranda Wei, Mainack Mondal, Euirim Choi, Shawn Shan, Claire Dolin, Michelle L. Mazurek, and Blase Ur. "Oh, the Places You've Been! User Reactions to Longitudinal Transparency About Third-Party Web Tracking and Inferencing." In *Proceedings of the 2019 ACM SIGSAC Conference on Computer and Communications Security*, 149–166. London United Kingdom: ACM, 2019. https://doi.org/10.1145/3319535.3363200.

The White House. "Executive Order on the Safe, Secure, and Trustworthy Development and Use of Artificial Intelligence." October 30, 2023. https://www.whitehouse.gov/briefing-room/presidential-actions/2023/10/30/executive-order-on-the-safe-secure-and-trustworthy-development-and-use-of-artificial-intelligence/.

Wiggers, Kyle. "A Timeline of Sam Altman's Firing from OpenAI—and the Fallout." *TechCrunch*, January 5, 2024. https://techcrunch.com/2024/01/05/a-timeline-of-sam-altmans-firing-from-openai-and-the-fallout/.

Wikimedia Foundation. "Our Work." Accessed July 29, 2023. https://wikimediafoundation.org/our-work/.

"Wikipedia: Editorial Oversight and Control." In *Wikipedia*, August 12, 2022. https://en.wikipedia.org/w/index.php?title=Wikipedia:Editorial_oversight_and_control.

Witynski, Max. "Perfect Pitch, Explained." *UChicago News*. Accessed September 18, 2023. https://news.uchicago.edu/explainer/what-is-perfect-pitch.

WNYC Studios. "Driverless Dilemma." *Radiolab Podcasts*, September 16, 2017. https://www.radiolab.org/episodes/driverless-dilemma.

Wood, Allen W. *Kant's Ethical Thought*. Modern European Philosophy. New York: Cambridge University Press, 1999.

World Bank. "Gains in Financial Inclusion, Gains for a Sustainable World." May 18, 2018. https://www.worldbank.org/en/news/immersive-story/2018/05/18/gains-in-financial-inclusion-gains-for-a-sustainable-world.

Wuerthele, Mike. "Apple Getting Sued over App Store User Data Collection." *AppleInsider*, November 12, 2022. https://appleinsider.com/articles/22/11/12/apple-getting-sued-over-app-store-user-data-collection.

Xiong, Jianghao, En-Lin Hsiang, Ziqian He, Tao Zhan, and Shin-Tson Wu. "Augmented Reality and Virtual Reality Displays: Emerging Technologies and Future Perspectives." *Light: Science & Applications* 10, no. 1 (October 25, 2021): 216. https://doi.org/10.1038/s41377-021-00658-8.

Yang, Grace S., Bryan Gibson, Adam K. Lueke, L. Rowell Huesmann, and Brad J. Bushman. "Effects of Avatar Race in Violent Video Games on Racial Attitudes and Aggression." *Social Psychological and Personality Science* 5, no. 6 (August 2014): 698–704. https://doi.org/10.1177/1948550614528008.

Zadrozny, Brandy. "Facebook to Crack down on Groups That Break Its Rules." *NBC News*, March 17, 2021. https://www.nbcnews.com/tech/tech-news/facebook-crack-groups-break-rules-rcna435.

Zhang, Sarah. "The Obscure Neuroscience Problem That's Plaguing VR." *Wired*, August 11, 2015. https://www.wired.com/2015/08/obscure-neuroscience-problem-thats-plaguing-vr/.

Zimmerli, Lukas, Mario Jacky, Lars Lünenburger, Robert Riener, and Marc Bolliger. "Increasing Patient Engagement During Virtual Reality-Based Motor Rehabilitation." *Archives of Physical Medicine and Rehabilitation* 94, no. 9 (September 2013): 1737–1746. https://doi.org/10.1016/j.apmr.2013.01.029.

Zuboff, Shoshana. *The Age of Surveillance Capitalism: The Fight for a Human Future at the New Frontier of Power*. New York: PublicAffairs, 2020.

Index

About the Author

Michael J. Grabowski, PhD, is professor of Communication, Sound, and Media Arts and director of the Game Design and Production program at Manhattan University. His work explores how forms of mediated communication, including immersive and AI-generated media, shape the way people think and act within their symbolic environment. He is the editor of the collection *Neuroscience and Media: New Understandings and Representations*, and his work on documentaries, feature films, commercials, music videos, and news has played at the Guggenheim, the Smithsonian, film festivals, and on broadcast, cable, and streaming television networks.

www.ingramcontent.com/pod-product-compliance
Lightning Source LLC
LaVergne TN
LVHW091422200226
832212LV00002B/5